Photodetectors: Technology and Applications

Photodetectors: Technology and Applications

Theodore Graves

WILLFORD PRESS

www.willfordpress.com

Published by Willford Press,
118-35 Queens Blvd., Suite 400,
Forest Hills, NY 11375, USA

ISBN: 978-1-64728-326-1

Cataloging-in-Publication Data

Photodetectors : technology and applications / Theodore Graves.
p. cm.
Includes bibliographical references and index.
ISBN 978-1-64728-326-1
1. Optical detectors. 2. Detectors. 3. Optical transducers. 4. Optoelectronic devices. I. Graves, Theodore.
TK8360.O67 P46 2022
681.25--dc23

For information on all Willford Press publications
visit our website at www.willfordpress.com

WILLFORD PRESS

Table of Contents

Preface

This book has been written, keeping in view that students want more practical information. Thus, my aim has been to make it as comprehensive as possible for the readers. I would like to extend my thanks to my family and co-workers for their knowledge, support and encouragement all along.

The sensors of light or other electromagnetic radiations are referred to as photodetectors. A photodetector has a p-n junction that converts the light photons into current. They can be broadly classified on the basis of their mechanism of detection into photoemission, thermal polarization and photochemical. Photoemission photodetectors use photoelectric effect where electrons transition from conduction bands of a material to free electron state. The most common examples of photodetectors are photodiodes and photo transistors. Some of the fundamental properties of photodectors are quantum efficiency, detectivity, dark current, responsivity, noise-equivalent power, nonlinearity and noise spectrum. Photodetectors include various devices such as gaseous ionization detectors, photomultiplier tubes, microchannnel plate detectors, active pixel sensors, charge-coupled devices, photoresistors, cryogenic detectors, etc. This book traces the progress of this field and highlights some of its key concepts. It is compiled in such a manner, that it will provide an in-depth knowledge about the technology and applications of this field. Scientists and students actively engaged in this field will find this book full of crucial and unexplored aspects of this domain.

A brief description of the chapters is provided below for further understanding:

Chapter – Introduction

Sensors that are used to detect light or any other electromagnetic radiation are referred to as photodetectors. They make use of p-n junction diode that converts light into electric current. Some of its types are phototransistors, photoconductive detectors, photomultipliers, etc. This is an introductory chapter which will briefly introduce all the significant aspects of photodetectors.

Chapter – Properties of Photodetectors

There are multiple properties that a photodetector possesses. Spectral response, quantum efficiency, responsivity, noise-equivalent power, detectivity, dark current, etc. are among its various properties. This chapter has been carefully written to provide an easy understanding of these varied properties of photodetectors.

Chapter – Photodiodes

A device in photodetectors which converts light radiations into electric current by absorbing photons is called photodiode. Types of photodiodes are p-n junction photodiode, P-i-N photodiode, avalanche photodiode and schottky photodiode. The topics elaborated in this chapter will help in gaining a better perspective about different types and aspects of photodiodes.

Chapter – Photovoltaic Cells

A specially designed electrical device that consists of semiconductor diode and converts light energy into direct current is termed as a photovoltaic cell. Its three major types are monocrystalline silicon cells, polycrystalline silicon cells and thin film cells. This chapter closely examines these types of photovoltaic cells and its related aspects to provide an extensive understanding of the subject.

Chapter – Phototransistors

An electronic component which is either used for current switching or current amplification is defined as phototransistor. It relies on light and operates in reverse bias connection. This chapter delves into the study of characteristics, applications and advantages and disadvantages of phototransistors to provide in-depth knowledge of the subject.

Chapter – Photoresistors

Photoresistor is a device that exhibits photoconductivity by decreasing the resistance with increasing intensity of light on its sensitive surface. It can be categorized into intrinsic photoresistor and extrinsic photoresistor. This chapter sheds light on these types of photoresistors and their applications for providing a thorough understanding of the subject.

Chapter – Thermal Photodetectors

Thermal photodetectors or biometers are sensors of light with a p-n junction to convert photons into current. Terahertz photodetectors, bolometers, microbolometer, pyroelectric detectors, etc. fall under its domain. This chapter delves into the subject of thermal photodetectors for an extensive understanding of it.

Chapter – Organic Photodetector

Organic photodetector consists of metal electrodes and organic small molecules as donors with fullerene derivatives as acceptors. Organic semiconductor and organic photodetector fabrication are some of its elements. This chapter has been carefully written to provide an easy understanding of organic photodetectors.

Chapter – Semiconductor Photodetectors

Semiconductor photodetectors consist of active-pixel sensor, cadmium zinc telluride, charge-coupled device, mercury cadmium telluride, light-emitting diode, quantum dot and silicon drift detector. This chapter closely examines about semiconductor photodetectors to provide an extensive understanding of the subject.

Chapter – Applications of Photodetectors

The varied applications of photodetectors lie in photoconductors, photomultiplier tubes, smoke detectors, compact disc players, televisions, remote controllers, microwave photonics, transmission media, etc. All these applications of photodetectors have been carefully analyzed in this chapter.

Theodore Graves

1

Introduction

Sensors that are used to detect light or any other electromagnetic radiation are referred to as photodetectors. They make use of p-n junction diode that converts light into electric current. Some of its types are phototransistors, photoconductive detectors, photomultipliers, etc. This is an introductory chapter which will briefly introduce all the significant aspects of photodetectors.

PHOTODETECTORS

Photodetectors are devices used for the detection of light – in most cases of optical powers. More specifically, photodetectors are usually understood as photon detectors, which in some way utilize the photo-excitation of electric carriers.

Photodetectors usually deliver an electronic output signal – for example, a voltage or electric current which is proportional to the incident optical power. They are thus belonging to the area of optoelectronics.

PHOTOELECTRIC EFFECT

Photoelectric effect is a phenomenon in which electrically charged particles are released from or within a material when it absorbs electromagnetic radiation. The effect is often defined as the ejection of electrons from a metal plate when light falls on it. In a broader definition, the radiant energy may be infrared, visible, or ultraviolet light, X rays, or gamma rays; the material may be a solid, liquid, or gas; and the released particles may be ions (electrically charged atoms or molecules) as well as electrons. The phenomenon was fundamentally significant in the development of modern physics because of the puzzling questions it raised about the nature of light—particle versus wavelike behaviour—that were finally resolved by Albert Einstein in 1905. The effect remains important for research in areas from materials science to astrophysics, as well as forming the basis for a variety of useful devices.

Photoelectric Principles

According to quantum mechanics, electrons bound to atoms occur in specific electronic configurations. The highest energy configuration (or energy band) that is normally occupied by electrons for a given material is known as the valence band, and the degree to which it is filled largely determines the material's electrical conductivity. In a typical conductor (metal), the valence band is about half filled with electrons, which readily move from atom to atom, carrying a current. In a good insulator, such as glass or rubber, the valence band is filled, and these valence electrons have very little mobility. Like insulators, semiconductors generally have their valence bands filled, but, unlike insulators, very little energy is required to excite an electron from the valence band to the next allowed energy band—known as the conduction band, because any electron excited to this higher energy level is relatively free. For example, the "bandgap" for silicon is 1.12 eV (electron volts), and that of gallium arsenide is 1.42 eV. This is in the range of energy carried by photons of infrared and visible light, which can therefore raise electrons in semiconductors to the conduction band. (For comparison, an ordinary flashlight battery imparts 1.5 eV to each electron that passes through it. Much more energetic radiation is required to overcome the bandgap in insulators.) Depending on how the semiconducting material is configured, this radiation may enhance its electrical conductivity by adding to an electric current already induced by an applied voltage, or it may generate a voltage independently of any external voltage sources.

Photoconductivity arises from the electrons freed by the light and from a flow of positive charge as well. Electrons raised to the conduction band correspond to missing negative charges in the valence band, called "holes." Both electrons and holes increase current flow when the semiconductor is illuminated.

In the photovoltaic effect, a voltage is generated when the electrons freed by the incident light are separated from the holes that are generated, producing a difference in electrical potential. This is typically done by using a p-n junction rather than a pure semiconductor. A p-n junction occurs at the juncture between p-type (positive) and n-type (negative) semiconductors. These opposite regions are created by the addition of different impurities to produce excess electrons (n-type) or excess holes (p-type). Illumination frees electrons and holes on opposite sides of the junction to produce a voltage across the junction that can propel current, thereby converting light into electrical power.

Other photoelectric effects are caused by radiation at higher frequencies, such as X rays and gamma rays. These higher-energy photons can even release electrons near the atomic nucleus, where they are tightly bound. When such an inner electron is ejected, a higher-energy outer electron quickly drops down to fill the vacancy. The excess energy results in the emission of one or more additional electrons from the atom, which is called the Auger effect.

Also seen at high photon energies is the Compton effect, which arises when an X-ray or gamma-ray photon collides with an electron. The effect can be analyzed by the same principles that govern the collision between any two bodies, including conservation of momentum. The photon loses energy to the electron, a decrease that corresponds to an increased photon wavelength according to Einstein's relation $E = hc/\lambda$. When the collision is such that the electron and the photon part at right angles to each other, the photon's wavelength increases by a characteristic amount called the Compton wavelength, 2.43×10^{-12} metre.

Applications

Devices based on the photoelectric effect have several desirable properties, including producing a current that is directly proportional to light intensity and a very fast response time. One basic device is the photoelectric cell, or photodiode. Originally, this was a phototube, a vacuum tube containing a cathode made of a metal with a small work function so that electrons would be easily emitted. The current released by the plate would be gathered by an anode held at a large positive voltage relative to the cathode. Phototubes have been replaced by semiconductor-based photodiodes that can detect light, measure its intensity, control other devices as a function of illumination, and turn light into electrical energy. These devices work at low voltages, comparable to their bandgaps, and they are used in industrial process control, pollution monitoring, light detection within fibre optics telecommunications networks, solar cells, imaging, and many other applications.

Photoconductive cells are made of semiconductors with bandgaps that correspond to the photon energies to be sensed. For example, photographic exposure meters and automatic switches for street lighting operate in the visible spectrum, so they are typically made of cadmium sulfide. Infrared detectors, such as sensors for night-vision applications, may be made of lead sulfide or mercury cadmium telluride.

Photovoltaic devices typically incorporate a semiconductor p-n junction. For solar cell use, they are usually made of crystalline silicon and convert about 15 percent of the incident light energy into electricity. Solar cells are often used to provide relatively small amounts of power in special environments such as space satellites and remote telephone installations. Development of cheaper materials and higher efficiencies may make solar power economically feasible for large-scale applications.

The photomultiplier tube is a highly sensitive extension of the phototube, first developed in the 1930s, which contains a series of metal plates called dynodes. Light striking the cathode releases electrons. These are attracted to the first dynode, where they release additional electrons that strike the second dynode, and so on. After up to 10 dynode stages, the photocurrent is so enormously amplified that some photomultipliers can virtually detect a single photon. These devices, or solid-state versions of comparable sensitivity, are invaluable in spectroscopy research, where it is often necessary to measure extremely weak light sources. They are also used in scintillation counters, which contain a material that produces flashes of light when struck by X rays or gamma rays, coupled to a photomultiplier that counts the flashes and measures their intensity. These counters support applications such as identifying particular isotopes for nuclear tracer analysis and detecting X rays used in computerized axial tomography (CAT) scans to portray a cross section through the body.

Photodiodes and photomultipliers also contribute to imaging technology. Light amplifiers or image intensifiers, television camera tubes, and image-storage tubes use the fact that the electron emission from each point on a cathode is determined by the number of photons arriving at that point. An optical image falling on one side of a semitransparent cathode is converted into an equivalent "electron current" image on the other side. Then electric and magnetic fields are used to focus the electrons onto a phosphor screen. Each electron striking the phosphor produces a flash of light, causing the release of many more electrons from the corresponding point on a cathode directly opposite the phosphor. The resulting intensified image can be further enhanced by the same process to produce even greater amplification and can be displayed or stored.

At higher photon energies the analysis of electrons emitted by X rays gives information about electronic transitions among energy states in atoms and molecules. It also contributes to the study of certain nuclear processes, and it plays a role in the chemical analysis of materials, since emitted electrons carry a specific energy that is characteristic of the atomic source. The Compton effect is also used to analyze the properties of materials, and in astronomy it is used to analyze gamma rays that come from cosmic sources.

TYPES OF PHOTODETECTORS

As the requirements for applications vary considerably, there are many types of photodetectors which may be appropriate in a particular case:

- Photodiodes are semiconductor devices with a p–n junction or p–i–n structure (i = intrinsic material) (→ p–i–n photodiodes), where light is absorbed in a depletion region and generates a photocurrent. Such devices can be very compact, fast, highly linear, and exhibit a high quantum efficiency (i.e., generate nearly one electron per incident photon) and a high dynamic range, provided that they are operated in combination with suitable electronics. A particularly sensitive type is that of avalanche photodiodes, which are sometimes used even for photon counting.

- Metal–semiconductor–metal (MSM) photodetectors contain two Schottky contacts instead of a p–n junction. They are potentially faster than photodiodes, with bandwidths up to hundreds of gigahertz.

- Phototransistors are similar to photodiodes, but exploit internal amplification of the photocurrent. They are less frequently used than photodiodes.

- Photoconductive detectors are also based on certain semiconductors, e.g. cadmium sulfide (CdS). They are cheaper than photodiodes, but they are fairly slow, are not very sensitive, and exhibit a nonlinear response. On the other hand, they can respond to long-wavelength infrared light.

- Phototubes are vacuum tubes or gas-filled tubes where the photoelectric effect is exploited ($\rightarrow$ photoemissive detectors).

- Photomultipliers are a special kind of phototubes, based on vacuum tubes. They can exhibit the combination of an extremely high sensitivity (even for photon counting) with a high speed and large active area. Some of them are based on multichannel plates; they can be substantially more compact than traditional photomultipliers.

- Research is performed on novel photodetectors based on carbon nanotubes (CNT) and graphene, which can offer a very broad wavelength range and a very fast response. Ways for integrating such devices into optoelectronic chips are explored.

These devices are all based on the internal or external photoelectric effect; photoemissive detectors belong to the latter category.

Various kinds of photodetectors can be integrated into devices like power meters and optical power monitors. Others can be made in the form of large two-dimensional arrays, e.g. for imaging applications. They may be called focal plane arrays. For example, there are CCD and CMOS sensors which are used mainly in cameras.

2

Properties of Photodetectors

There are multiple properties that a photodetector possesses. Spectral response, quantum efficiency, responsivity, noise-equivalent power, detectivity, dark current, etc. are among its various properties. This chapter has been carefully written to provide an easy understanding of these varied properties of photodetectors.

SPECTRAL RESPONSE

The spectral response of a photodetector is the range of optical wavelengths or frequencies in which the detector has a significant responsivity. There is no universally defined criterion for the minimum responsivity; it may, for example, be taken as one tenth of the maximum responsivity, or even much less. In other cases, a drop of at most 50% may be acceptable for an application. Due to that uncertainty, given specifications can vary even for the same device.

Note that the spectral range quoted for an optical power meter, for example, may be smaller than the spectral response: there may be wavelength regions where the detector reacts, but not with a calibrated response.

Typical Limiting Factors

Some typical limiting factors for the spectral response of photodetectors are:

- Many types of detectors, for example all photoemissive detectors and all semiconductor-based detectors containing a p–n junction, work only for photon energies above a certain level. That condition translates into some maximum optical wavelength.

- Such a limitation does not occur for thermal detectors, which therefore can exhibit a very broad and smooth spectral response. There, however, the spectral response may be limited by the wavelength-dependent absorption of the used absorber.

- Many detectors have an optical window, e.g. for protecting the light-sensitive area or for preserving a vacuum inside the detector, and that window has a limited wavelength range with high transmissivity.

- Sometimes, the spectral response of a photodetector is intentionally limited with an optical filter, because a response to certain other wavelengths is undesirable for a particular application. In some cases, one wants to avoid degradation effects caused by short-wavelength light (e.g. ultraviolet light).

For a given material, e.g. of a photocathode, the spectral response may substantially vary due to different factors, e.g. the applied thickness of a layer, an additional reflector or details of the material fabrication process.

QUANTUM EFFICIENCY

The term quantum efficiency (QE) may apply to incident photon to converted electron (IPCE) ratio, of a photosensitive device or it may refer to the TMR effect of a Magnetic Tunnel Junction.

In a charge-coupled device (CCD) it is the percentage of photons hitting the device's photoreactive surface that produce charge carriers. It is measured in electrons per photon or amps per watt. Since the energy of a photon is inversely proportional to its wavelength, QE is often measured over a range of different wavelengths to characterize a device's efficiency at each photon energy level. The QE for photons with energy below the band gap is zero. Photographic film typically has a QE of much less than 10%, while CCDs can have a QE of well over 90% at some wavelengths.

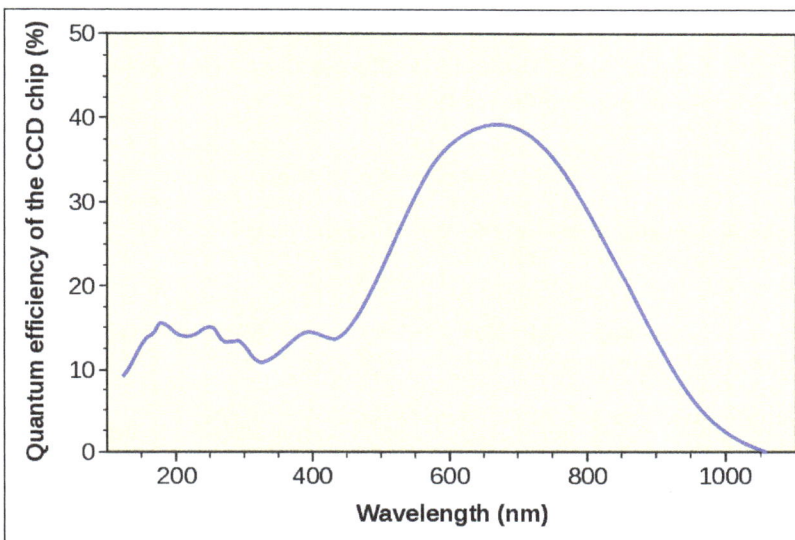

A graph showing variation of quantum efficiency with wavelength of a CCD chip in the Hubble Space Telescope's Wide Field and Planetary Camera 3.

Solar Cells

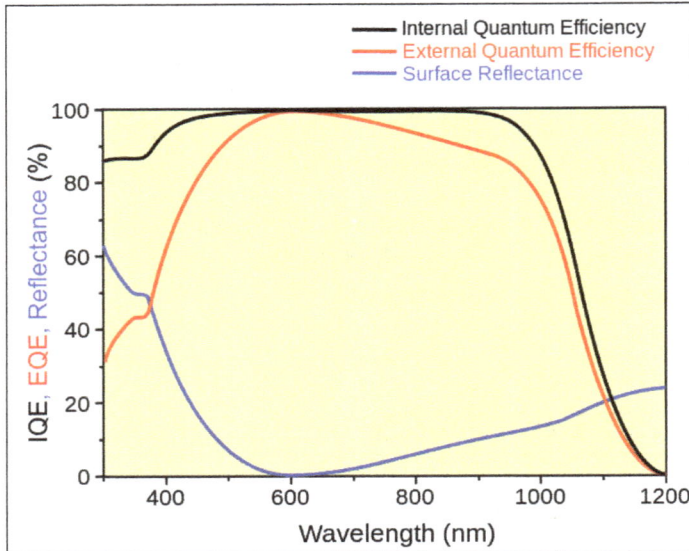

A graph showing variation of internal quantum efficiency, external quantum efficiency, and reflectance with wavelength of a crystalline silicon solar cell.

A solar cell's quantum efficiency value indicates the amount of current that the cell will produce when irradiated by photons of a particular wavelength. If the cell's quantum efficiency is integrated over the whole solar electromagnetic spectrum, one can evaluate the amount of current that the cell will produce when exposed to sunlight. The ratio between this energy-production value and the highest possible energy-production value for the cell (i.e., if the QE were 100% over the whole spectrum) gives the cell's overall energy conversion efficiency value. Note that in the event of multiple exciton generation (MEG), quantum efficiencies of greater than 100% may be achieved since the incident photons have more than twice the band gap energy and can create two or more electron-hole pairs per incident photon.

Types

Two types of quantum efficiency of a solar cell are often considered:

- External Quantum Efficiency (EQE) is the ratio of the number of charge carriers collected by the solar cell to the number of photons of a given energy shining on the solar cell from outside (incident photons).

- Internal Quantum Efficiency (IQE) is the ratio of the number of charge carriers collected by the solar cell to the number of photons of a given energy that shine on the solar cell from outside and are absorbed by the cell.

The IQE is always larger than the EQE. A low IQE indicates that the active layer of the solar cell is unable to make good use of the photons. To measure the IQE, one first measures the EQE of the solar device, then measures its transmission and reflection, and combines these data to infer the IQE.

$$EQE = \frac{electrons/sec}{photons/sec} = \frac{(current)/(charge\,of\,one\,electron)}{(total\,power\,of\,photons)/(energy\,of\,one\,photon)}$$

$$IQE = \frac{electrons/sec}{absorbed\,photons/sec} = \frac{EQE}{1\text{-}Reflection\text{-}Transmission}$$

The external quantum efficiency therefore depends on both the absorption of light and the collection of charges. Once a photon has been absorbed and has generated an electron-hole pair, these charges must be separated and collected at the junction. A "good" material avoids charge recombination. Charge recombination causes a drop in the external quantum efficiency.

The ideal quantum efficiency graph has a square shape, where the QE value is fairly constant across the entire spectrum of wavelengths measured. However, the QE for most solar cells is reduced because of the effects of recombination, where charge carriers are not able to move into an external circuit. The same mechanisms that affect the collection probability also affect the QE. For example, modifying the front surface can affect carriers generated near the surface. And because high-energy (blue) light is absorbed very close to the surface, considerable recombination at the front surface will affect the "blue" portion of the QE. Similarly, lower energy (green) light is absorbed in the bulk of a solar cell, and a low diffusion length will affect the collection probability from the solar cell bulk, reducing the QE in the green portion of the spectrum. Generally, solar cells on the market today do not produce much electricity from ultraviolet and infrared light (<400 nm and >1100 nm wavelengths, respectively); these wavelengths of light are either filtered out or are absorbed by the cell, thus heating the cell. That heat is wasted energy, and could damage the cell.

Quantum Efficiency of Image Sensors: Quantum efficiency (QE) is the fraction of photon flux that contributes to the photocurrent in a photodetector or a pixel. Quantum efficiency is one of the most important parameters used to evaluate the quality of a detector and is often called the spectral response to reflect its wavelength dependence. It is defined as the number of signal electrons created per incident photon. In some cases it can exceed 100% (i.e. when more than one electron is created per incident photon).

EQE Mapping: Conventional measurement of the EQE will give the efficiency of the overall device. However it is often useful to have a map of the EQE over large area of the device. This mapping provides an efficient way to visualize the homogeneity and/or the defects in the sample. It was realized by researchers from the Institute of Researcher and Development on Photovoltaic Energy (IRDEP) who calculated the EQE mapping from electroluminescence measurements taken with a hyperspectral imager.

Spectral Responsivity

Spectral responsivity is a similar measurement, but it has different units: amperes per watt (A/W); (i.e. how much current comes out of the device per incoming photon of a given energy and wavelength). Both the quantum efficiency and the responsivity are functions of the photons' wavelength (indicated by the subscript λ).

To convert from responsivity (R_λ, in A/W) to QE_λ (on a scale 0 to 1):

$$QE_\lambda = \frac{R_\lambda}{\lambda} \times \frac{hc}{e} \approx \frac{R_\lambda}{\lambda} \times (1240\,\text{W}\cdot\text{nm/A})$$

where λ is the wavelength in nm, h is the Planck constant, c is the speed of light in a vacuum, and e is the elementary charge.

Determination

$$QE_\lambda = \eta = \frac{N_e}{N_v}$$

where Φ_ξ = number of electrons produced, N_v = number of photons absorbed.

$$\frac{N_v}{t} = \Phi_o \frac{\lambda}{hc}$$

Assuming each photon absorbed in the depletion layer produces a viable electron-hole pair, and all other photons do not,

$$\frac{N_e}{t} = \Phi_\xi \frac{\lambda}{hc}$$

where t is the measurement time (in seconds), Φ_o = incident optical power in watts, Φ_ξ = optical power absorbed in depletion layer, also in watts.

RESPONSIVITY

Responsivity measures the input–output gain of a detector system. In the specific case of a photodetector, responsivity measures the electrical output per optical input.

The responsivity of a photodetector is usually expressed in units of either amperes or volts per watt of incident radiant power. For a system that responds linearly to its input, there is a unique responsivity. For nonlinear systems, the responsivity is the local slope. Many common photodetectors respond linearly as a function of the incident power.

Responsivity is a function of the wavelength of the incident radiation and of the sensor properties, such as the bandgap of the material of which the photodetector is made. One simple expression for the responsivity R of a photodetector in which an optical signal is converted into an electric current (known as a photocurrent) is:

$$R = \eta \frac{q}{hf} \approx \eta \frac{\lambda_{(\mu m)}}{1.23985(\mu m \times W / A)}$$

where η is the quantum efficiency (the conversion efficiency of photons to electrons) of the detector for a given wavelength, q is the electron charge, f is the frequency of the optical signal, and h is Planck's constant. This expression is also given in terms of λ, the wavelength of the optical signal, and has units of amperes per watt (A/W).

The term responsivity is also used to summarize input–output relationship in non-electrical systems. For example, a neuroscientist may measure how neurons in the visual pathway respond to light. In this case, responsivity summarizes the change in the neural response per unit signal strength. The responsivity in these applications can have a variety of units. The signal strength typically is controlled by varying either intensity (intensity-response function) or contrast (contrast-response function). The neural response measure depends on the part of the nervous system under study. For example, at the level of the retinal cones, the response might be in photocurrent. In the central nervous system the response is usually spikes per second. In functional neuroimaging, the response measure is usually BOLD contrast. The responsivity units reflect the relevant stimulus and physiological units.

When describing an amplifier, the more common term is gain. Deprecated synonym sensitivity. A system's sensitivity is the inverse of the stimulus level required to produce a threshold response, with the threshold typically chosen just above the noise level.

NOISE-EQUIVALENT POWER

Noise-equivalent power (NEP) is a measure of the sensitivity of a photodetector or detector system. It is defined as the signal power that gives a signal-to-noise ratio of one in a one hertz output bandwidth. An output bandwidth of one hertz is equivalent to half a second of integration time. The units of NEP are watts per square root hertz. The NEP is equal to the noise spectral density (expressed in units of $A/\sqrt{Hz}$ or $V/\sqrt{Hz}$) divided by the responsivity (expressed in units of A/W or V/W, respectively).

A smaller NEP corresponds to a more sensitive detector. For example, a detector with an NEP of $10^{-12} W/\sqrt{Hz}$ can detect a signal power of one picowatt with a signal-to-noise ratio (SNR) of one after one half second of averaging. The SNR improves as the square root of the averaging time, and hence the SNR in this example can be improved by a factor of 10 by averaging 100-times longer, i.e. for 50 seconds.

If the NEP refers to the signal power absorbed in the detector, it is known as the electrical NEP. If instead it refers to the signal power incident on the detector system, it is called the optical NEP. The optical NEP is equal to the electrical NEP divided by the optical coupling efficiency of the detector system.

DETECTIVITY

The detectivity D of a photodetector is a figure of merit, defined as the inverse of the noise-equivalent power (NEP). The larger the detectivity of a photodetector, the more it is suitable for detecting week signals which compete with the detector noise.

The specific detectivity D^* is the detectivity normalized to a unit detector area and detection bandwidth; one can calculate it by multiplying the detectivity with the square root of the product of detector area (in square centimeters) and the detector bandwidth (in Hz). That term is useful for comparing the performance of different detector technologies. If the detector bandwidth scales inversely with the active area, e.g. because of the limiting impact of the electrical capacitance, the specific detectivity will be independent of the active area.

For a given detector, one may decrease the noise level, thus reducing the noise-equivalent power and increasing the detectivity, by restricting its detection bandwidth (e.g. by adding a low-pass filter). Assuming white noise, one will then obtain the same specific detectivity as before.

DARK CURRENT

Most photodetectors such as photodiodes, phototransistors, CCD sensors and phototubes produce a signal current which is more or less proportional to the incident optical power. However, even in the absence of any light input, there is often some tiny amount of DC current, which one calls the dark current. An also possible fluctuating thermal current with zero mean value is usually not called a dark current.

For many applications, the dark current is totally negligible, but in some cases it matters – for example, when extremely small optical powers need to be detected. One may in principle subtract the dark current from the obtained signal either with analog electronics or with software, but that works only to a limited extent, because the dark current can be substantially temperature-dependent, and it also exhibits shot noise.

Origins of Dark Current

Dark Current in Photodetectors with Internal Photoelectric Effect

In photodiodes and other detectors with some p–n or p–i–n junction, it is often caused by thermal excitation (generation) of carriers – not necessarily directly from valence to conduction band, but possibly through defect states related to crystal defects or impurities. The rate of such thermal processes depends not only on the active area, but also critically on the temperature and on the band gap energy of the material, and also on the operation voltage (particularly near the breakdown voltage, where impact ionization can occur). At high voltages, tunneling through the depletion region may also contribute.

For visible light detectors such as silicon-based photodiodes, the dark current can be very small (e.g. in the picoampere region) (even for significant bias voltages) and is then negligible for most applications. Germanium photodiodes exhibit much higher dark currents which is however mostly not due to their somewhat lower band energy. Indium gallium arsenide diodes, which also have a reduced bandgap energy compared with silicon, also exhibit a relatively low dark current.

For materials with substantially smaller band gap, dark current can be a serious problem and may thus enforce the operation at substantially reduced temperatures. Therefore, some mid-infrared cameras, for example, need to be equipped with a Stirling cooler for operation around 100 K or even lower.

For operation near the break-down voltage, the dark current can become far stronger than for lower voltages. Dark currents may also be generated by some leakage currents which are not related to thermal excitation.

In any case, a dark current can normally not occur for operation with zero bias voltage, since there is no energy supply available for it – at least as long as the temperature of the device is uniform, excluding any Peltier effects. Therefore, one may operate a photodiode, for example, with zero bias voltage in cases where influences of a dark current must be avoided.

Of course, drifts of output signals may also occur in related electronics, for example due to bias drifts of operational amplifiers. Therefore, a non-zero output signal does not necessarily indicate a dark current of the detector.

Dark Current in Photodetectors with External Photoelectric Effect

The primary cause for a dark current is usually thermionic emission on the photocathode. This means the thermal excitation of electrons. Thermionic emission can be substantial for cathode materials with very low work function, as required for infrared detection. It is also strongly temperature-dependent; low-temperature operation is thus a very effective measure for reducing the dark current. The dependence on the operation voltage is weak.

For quite high operation voltages, there can be a steeper rise of dark current due to field emission at various locations in the bulb. That can lead to accelerated aging.

Some current is contributed by the ionization of residual gas, i.e., due to the non-perfect vacuum. This is particularly the case for devices operated with higher voltages, for example photomultipliers.

A typically quite weak contribution comes from the leakage current due to non-perfect electrical isolation.

It is also possible that some unwanted light is generated by scintillation, e.g. when electrons hit the glass tube. At a usually very low level, there are weak flashes of light caused by cosmic rays and radioactive substances e.g. in the glass tube or the near surroundings.

References

- Spectral-response-of-a-photodetector: rp-photonics.com, Retrieved 14 February, 2019

- Detectivity: rp-photonics.com, Retrieved 10 March, 2019

- A. Delamarre; et al. (2014). "Quantitative luminescence mapping of Cu(In,Ga)Se2 thin-film solar cells". Progress in Photovoltaics. doi:10.1002/pip.2555

- Dark-current: rp-photonics.com, Retrieved 30 April, 2019

3

Photodiodes

A device in photodetectors which converts light radiations into electric current by absorbing photons is called photodiode. Types of photodiodes are p-n junction photodiode, P-i-N photodiode, avalanche photodiode and schottky photodiode. The topics elaborated in this chapter will help in gaining a better perspective about different types and aspects of photodiodes.

A photodiode is a p-n junction or pin semiconductor device that consumes light energy to generate electric current. It is also sometimes referred as photo-detector, photo-sensor, or light detector.

Photodiodes are specially designed to operate in reverse bias condition. Reverse bias means that the p-side of the photodiode is connected to the negative terminal of the battery and n-side is connected to the positive terminal of the battery.

Photodiode is very sensitive to light so when light or photons falls on the photodiode it easily converts light into electric current. Solar cell is also known as large area photodiode because it converts solar energy or light energy into electric energy. However, solar cell works only at bright light.

The construction and working of photodiode is almost similar to the normal p-n junction diode. PIN (p-type, intrinsic and n-type) structure is mostly used for constructing the photodiode instead of p-n (p-type and n-type) junction structure because PIN structure provide fast response time. PIN photodiodes are mostly used in high-speed applications.

In a normal p-n junction diode, voltage is used as the energy source to generate electric current whereas in photodiodes, both voltage and light are used as energy source to generate electric current.

Photodiode Symbol

The symbol of photodiode is similar to the normal p-n junction diode except that it contains arrows striking the diode. The arrows striking the diode represent light or photons.

A photodiode has two terminals: a cathode and an anode.

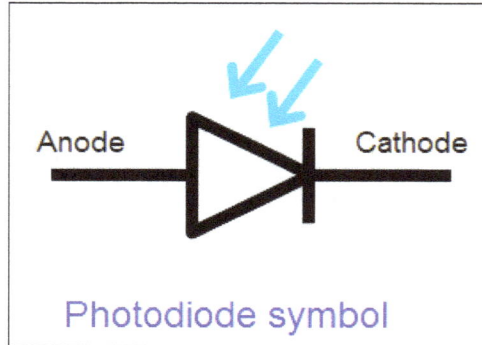

Photodiode symbol

Objectives and Limitations of Photodiode

- Photodiode should be always operated in reverse bias condition.

- Applied reverse bias voltage should be low.

- Generate low noise.

- High gain.

- High response speed.

- High sensitivity to light.

- Low sensitivity to temperature.

- Low cost.

- Small size.

- Long lifetime.

CONSTRUCTION OF PHOTODIODE

The photodiode is made up of two layers of P-type and N-type semiconductor. In this, the P-type material is formed from diffusion of the lightly doped P-type substrate. Thus, the layer of P+ ions is formed due to the diffusion process. And N-type epitaxial layer is grown on N-type substrate. The P+ diffusion layer is developed on N-type heavily doped epitaxial layer. The contacts are made up of metals to form two terminal cathode and anode.

The front area of the diode is divided into two types that are active surface and non-active surface. The non-active surface is made up of SiO_2 (Silicon di Oxide) and the active surface is coated with anti-reflection material. The active surface is called so because the light rays are incident on it.

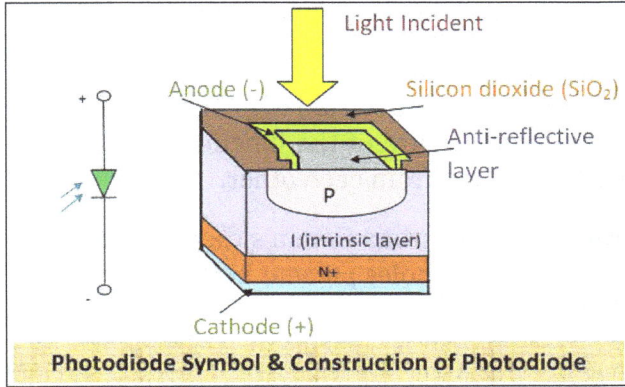

Photodiode Symbol & Construction of Photodiode

While on the non-active surface the light rays do not strike. The active layer is coated with anti-reflection material so that the light energy is not lost and the maximum of it can be converted into current. The entire unit has dimensions of the order of 2.5 mm.

WORKING OF PHOTODIODE

In the photodiode, a very small reverse current flows through the device that is termed as dark current. It is called so because this current is totally the result of the flow of minority carriers and is thus flows when the device is not exposed to radiation.

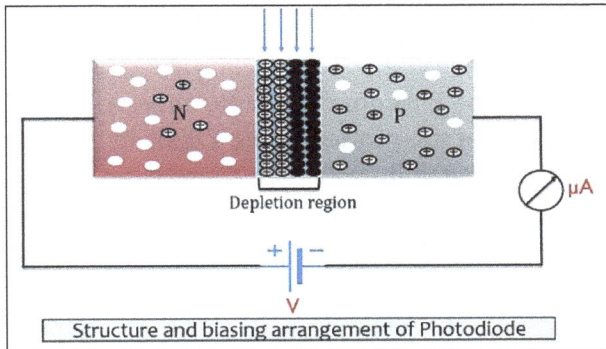

Structure and biasing arrangement of Photodiode

The electrons present in the p side and holes present in n side are the minority carriers. When a certain reverse-biased voltage is applied then minority carrier, holes from n-side experiences repulsive force from the positive potential of the battery.

Similarly, the electrons present in the p side experience repulsion from the negative potential of the battery. Due to this movement electron and hole recombine at the junction resultantly generating depletion region at the junction.

Due to this movement, a very small reverse current flows through the device known as dark current.

The combination of electron and hole at the junction generates neutral atom at the depletion. Due to which any further flow of current is restricted.

Now, the junction of the device is illuminated with light. As the light falls on the surface of the junction, then the temperature of the junction gets increased. This causes the electron and hole to get separated from each other.

At the two gets separated then electrons from n side gets attracted towards the positive potential of the battery. Similarly, holes present in the p side get attracted to the negative potential of the battery.

This movement then generates high reverse current through the device.

With the rise in the light intensity, more charge carriers are generated and flow through the device. Thereby, producing a large electric current through the device.

This current is then used to drive other circuits of the system.

So, we can say the intensity of light energy is directly proportional to the current through the device.

Only positive biased potential can put the device in no current condition in case of the photodiode.

CHARACTERISTICS OF PHOTODIODE

Photodiode operates in reverse bias condition. Reverse voltages are plotted along X axis in volts and reverse current are plotted along Y-axis in microampere. Reverse current does not depend on reverse voltage. When there is no light illumination, reverse current will be almost zero. The minimum amount of current present is called as Dark Current. Once when the light illumination increases, reverse current also increases linearly.

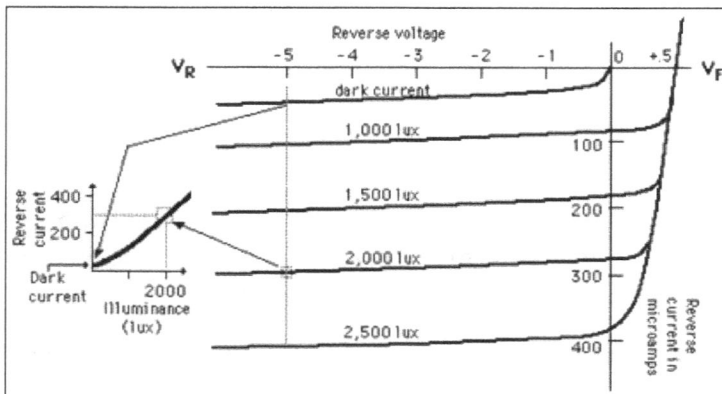

TYPES OF PHOTODIODE

P-i-N Junction Photodiodes

P-i-N photodiodes are commonly used in a variety of applications. A typical P-i-N pho-todiode is shown in figure. It consists of a highly-doped transparent p-type contact layer on top of an undoped absorbing layer and an n-type highly doped contact layer on the bottom. Discrete photodiodes are fabricated on a conductive substrate as shown in the figure, which facilitates the formation of the n-type contact and reduces the number of process steps. The top contact is typically a metal ring contact, which has a low contact resistance and still allows the light to be absorbed in the semiconductor. An alternative approach uses a transparent conductor such as Indium Tin Oxide (ITO). The active device area is formed by mesa etching or by proton implantation of the adjacent area, which makes it isolating. A dielectric layer is added around the active area to reduce leakage currents and to ensure a low parasitic capacitance of the contact pad.

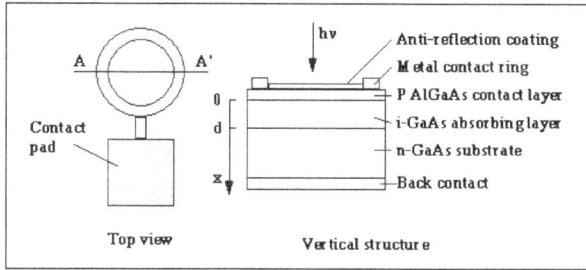

Top view and vertical structure through section A-A' of a P-i-N heterostructure photodiode.

Grading of the material composition between the transparent contact layer and the absorbing layer is commonly used to reduce the $n - n^+$ or $p - p^+$ barrier formed at the interface.

The above structure evolved mainly from one basic requirement: light should be absorbed in the depletion region of the diode to ensure that the electrons and holes are separated in the electric field and contribute to the photocurrent, while the transit time must be minimal.

This implies that a depletion region larger than the absorption length must exist in the detector. This is easily assured by making the absorbing layer undoped. Only a very small voltage is required to deplete the undoped region. If a minimum electric field is required throughout the absorbing layer, to ensure a short transit time, it is also the undoped structure, which satisfies this condition with a minimal voltage across the region, because the electric field is constant. An added advantage is that the recombination/generation time constant is longest for undoped material, which provides a minimal thermal generation current.

It also implies that the top contact layer should be transparent to the incoming light. In silicon photodiodes one uses a thin highly doped contact layer to minimize the absorption. By using a contact layer with a wider band gap (also called the window layer) absorption in the contact layer can be eliminated (except for a small fraction due to free carrier absorption) which improves the responsivity.

Electron-hole pairs, which are absorbed in the quasi-neutral regions, can still contribute to the photocurrent provided they are generated within one diffusion length of the depletion region.

However, the collection of carriers due to diffusion is relatively inefficient and leads to long tails in the transient response. It therefore should be avoided.

Because of the large difference in refractive index between air and most semiconductors, there is a substantial reflection at the surface. The reflection at normal incidence between two materials with refractive index n_1 and n_2 is given by:

$$R = (\frac{n_1 - n_2}{n_1 + n_2})^2$$

For instance, the reflection between air and GaAs (n = 3.5) is 31 %.

By coating the semiconductor surface with a dielectric material (anti-reflection coating) of appropriate thickness this reflection can largely be eliminated.

The reflectivity for an arbitrary incident angle is:

$$R_{TE} = (\frac{n_1 \cos\theta_i - n_2 \cos\theta_t}{n_1 \cos\theta_i + n_2 \cos\theta})^2$$

$$R_{TM} = (\frac{n_2 \cos\theta_i - n_1 \cos\theta_t}{n_2 \cos\theta_i + n_1 \cos\theta})^2$$

with $n_2 \sin\theta_t = n_t \sin\theta_i$

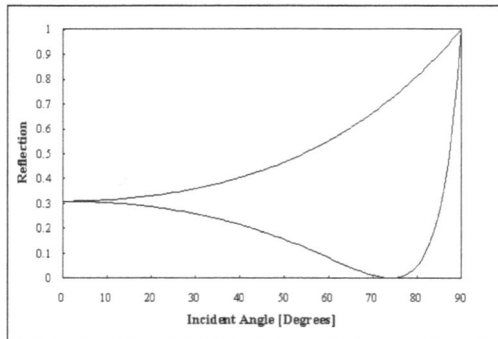

Reflectivity versus incident angle for a transverse electric, R_{TE}, and transverse magnetic, R_{TM}, incident field.

where θ_i is the incident angle, and θ_t the transmitted angle. $_{TE}$ is the reflectivity if the electric field is parallel to the surface while R_{TM} is the reflectivity if the magnetic field is parallel to the surface. The reflectivity as a function of θ_i, for an air-GaAs interface is shown in the figure.

Responsivity of a P-i-N Photodiode

Generation of Electron Hole Pairs

The generation of electron-hole pairs in a semiconductor is directly related to the absorption of light since every absorbed photon generates one electron-hole pair. The optical generation rate g_{op} is given by:

$$g_{op} = -\frac{1}{A}\frac{dP_{opt}}{dx}\frac{1}{h\upsilon} = \frac{\alpha P_{opt}}{Ah\upsilon}$$

where A is the illuminated area of the photodiode, P_{opt} is the incident optical power, α is the absorption coefficient and $h\upsilon$ is the photon energy. Note that the optical power is position dependent and obtained by solving:

$$\frac{dP_{opt}}{dx} = -\alpha P_{opt}$$

The resulting generation rate must be added to the continuity equation and solved throughout the photodiode, which results in the photocurrent.

Photocurrent due to Absorption in the Depletion Region

Assuming that all the generated electron-hole pairs contribute to the photocurrent, the photocurrent is simply the integral of the generation rate over the depletion region:

$$I_{ph} = -qA\int_{-x_p}^{x_n+d} g_{op}dx$$

where d is the thickness of the undoped region. The minus sign is due to the sign convention indicated on the figure. For a p-i-n diode with heavily doped n-type and p-type regions and a transparent top contact layer, this integral reduces to:

$$I_{ph} = -\frac{q(1-R)P_{in}}{h\upsilon}(1-e^{-\alpha d})$$

where P_{in} is the incident optical power and R is the reflection at the surface.

Photocurrent due to Absorption in the Quasi-neutral Region

To find the photocurrent due to absorption in the quasi-neutral region, we first have to solve the diffusion equation in the presence of light. For holes in the n-type contact layer this means solving the continuity equation:

$$\frac{\partial p_n}{dt} = -\frac{1}{q}\frac{\partial J_p}{dx} + \frac{p_n - p_{n0}}{\tau_p} + g_{op}(x)$$

Where the electron-hole pair generation g_{op} depends on position. For an the n-type contact layer with the same energy bandgap as the absorption layer, the optical generation rate equals:

$$g_{op}(x) = \frac{P_{in}(1-R)\alpha e^{-\alpha x}}{Ah\upsilon}$$

and the photocurrent due to holes originating in the n-type contact layer equals:

$$I_{ph} = -\frac{q(1-R)P_{in}e^{-\alpha d}}{h\upsilon}\frac{\alpha L_p}{1+\alpha L_p}(1-e^{-\alpha d}) - \frac{qD_p p_{n0}}{L_p}$$

The first term is due to light whereas the second term is the due to thermal generation of electron-hole pairs. This derivation assumes that the thickness of the n-type contact layer is much larger than the diffusion length.

Absorption in the P-contact Region

Even though the contact layer was designed so that no light absorbs in this layer, it will become absorbing at shorter wavelengths. Consider a worst-case scenario where all the electron-hole pairs, which are generated in the p-type contact layer, recombine without contributing to the photocurrent. The optical power incident on the undoped region is reduced by $\exp(\alpha^* w_p)$ where w_p is the width of the quasi-neutral p region and α^* is the absorption coefficient in that region.

Total Responsivity

Combining all the above effects the total responsivity of the detector - ignoring the dark current equals:

$$\mathcal{R} = \left|\frac{I_{ph}}{P_{in}}\right| = \frac{q(1-R)\exp(-a^* w_p')}{h\upsilon}[1 - \frac{e^{-\alpha d}}{1+\alpha L}]$$

Note that α^*, α and hn are wavelength dependent. For a direct bandgap semiconductor these are calculated from:

$$\alpha = K\sqrt{E_{ph} - E_g} \text{ and } \alpha^* = K^*\sqrt{E_{ph} - E_g^*}$$

The quantum efficiency then equals:

$$\eta = \frac{\mathcal{R}h\upsilon}{q} = (1-R)\exp(-a^* w'_p)[1 - \frac{e^{-\alpha d}}{1+\alpha L}]$$

Dark Current of the Photodiode

The dark current of a p-n diode including the ideal diode current, as well as recombination/generation in the depletion region is given by:

$$I_{p-i-n} = qA(\frac{D_p p_{n0}}{L_p} + \frac{D_n np0}{w'_p} + bdn_i^2, u)(e^{Va/V_t} - 1)$$

$$+ qA\frac{x' n_{i,u}}{2\tau_{nr}}(e^{V_a/2V_t} - 1).$$

Under reverse bias conditions this expression reduces to:

$$I_{p-i-n} = -qA[\frac{D_p p_{n0}}{L_p} + bdn_{i,u}^2 + \frac{x' n_{i,u}}{2\tau_{nr}}]$$

The ideal diode current due to recombination of electrons has been ignored since $n_{p0} = n_{i,p}^2 / N_a$ is much smaller than p_{n0} because the p-layer has a larger band gap. In the undoped region, one expects the trap-assisted generation to be much larger than bimolecular generation. Which further reduces the current to:

$$I_{p-i-n} = -qA[\frac{D_p n_{i,n}^2}{L_p N_d} + \frac{x' n_{i,u}}{2\tau_{nr}}]$$

The trap-assisted recombination tends to dominate for most practical diodes.

Noise in a Photodiode

Shot Noise Sensitivity

Noise in a p-i-n photodiode is primarily due to shot noise; the random nature of the generation of carriers in the photodiode yields also a random current fluctuation. The square of the current fluctuations equals:

$$<i^2> = 2q(\sum_j I_j)\Delta f$$

where I_j are the currents due to different recombination/generation mechanisms and Δf is the frequency range. Including the ideal diode current, Shockley-Hall-Read and band-to-band recombination as well as generation due to light one obtains:

$$< i^2 >= 2q\Delta f \{qA[\frac{D_n n_{p0}}{w'_p} + \frac{D_p p_{n0}}{L_p} + bd n_{i,u}^2 + \frac{x' n_{i,u}}{2\tau_{nr}}]$$

$$+ \frac{qAn_i x'}{2\tau_{nr}}(1 + e^{V_a/2V_t}) + P_{in}\mathcal{R}\}$$

The minimum detectable input power depends on the actual signal and the required signal to noise ration. As a first approximation, we now calculate the minimum detectable power as the power, which generates a current equal to the RMS noise current.

$$P_{min} = \frac{\sqrt{< i^2 >}}{\mathcal{R}}$$

The minimal noise current is obtained at $V_a = 0$ for which the noise current and minimal power equal:

$$< i^2 >= 2q \, \Delta f [2\frac{n_{i,n}^2 D_p}{L_p N_d} + 2\frac{n_{i,u} x'}{2\tau_{nr}} + P_{in}\mathcal{R}]$$

$$P_{min} = \sqrt{\frac{2q \, \Delta f}{\mathcal{R}^2}[2\frac{n_{i,n}^2 D_p}{L_p N_d} + 2\frac{n_{i,u} x'}{2\tau_{nr}} + P_{in}\mathcal{R}]}$$

Equivalence of Shot Noise and Johnson Noise

The following derivation illustrates that shot noise and Johnson noise are not two independent noise mechanisms. In fact, we will show that both are the same for the special case of an ideal pn diode under zero bias. At zero bias the photodetector can also be modeled as a resistor. Therefore the expression for Johnson noise should apply:

$$< i^2 >= \frac{4kT\Delta f}{R}$$

The resistance of a photodiode with $I = I_s(e^{V_a/V_t} - 1)$ is:

$$R = \frac{1}{\frac{dI}{dV_a}} = \frac{V_t}{I_s e^{V_a/V_t}}$$

or for zero bias, the Johnson noise current is given by:

$$< i^2 >= \frac{4kT\Delta f I_s}{V_t} = 4q I_s \, \Delta f$$

whereas the shot noise current at Va = 0 is given by:

$$<i^2> = 2qI_s(1 + e^{V_a/V_t}) = 4qI_s \Delta f$$

Where we added the noise due to the diffusion current to the noise due to the (constant) drift current, since both noise mechanisms do not cancel each other. Equations are identical, thereby proving the equivalence between shot noise and Johnson noise in a photodiode at zero voltage. Note that this relation does not apply if the current is dominated by trap-assisted recombination/generation in the depletion region because of the non-equilibrium nature of the recombination/generation process.

Examples

For a diode current of $1\mu A$, a bandwidth Δf of 1 GHz and a responsivity, $\mathcal{R}$, of 0.2A/W, the noise current $\sqrt{<i^2>}$ equals 18 nA, corresponding to a minimum detectable power of 89 nW or -40.5 dBm. Johnson noise in a 50Ω resistor, over a bandwidth Δf of 1 GHz, yields a noise current of $0.58\mu A$ and $P_{min} = 2.9\mu W$ or -25.4 dBm.

If the diode current is only due to the optical power, or $I = P_{min}\mathcal{R}$, then:

$$P_{min} = \frac{2q\Delta f}{\mathcal{R}} = -58\,\text{dBm}$$

The sensitivity for a given bandwidth can also be expressed as a number of photons per bit:

$$\#\text{photons/bit} = \frac{P_{min}}{h\upsilon\Delta f}$$

For instance, for a minimal power of -30 dBm and a bandwidth of 1 GHz, this sensitivity corresponds to 4400 photons per bit.

Noise Equivalent Power and AC Noise Analysis

Assume the optical power with average value P_0 is amplitude modulated with modulation depth, m, as described by:

$$P_{in} = P_0(1 + me^{j\omega t})$$

The ac current (RMS value) in the photodiode with responsivity, $\mathcal{R}$, is then:

$$i_{ph} = \frac{mP_0}{\sqrt{2}}\mathcal{R}$$

which yields as an equivalent circuit of the photodiode a current source $\sqrt{<i^2>}$ in parallel with a resistance, R_{eq}, where R_{eq} is the equivalent resistance across the diode and $\sqrt{<i^2>}$ is the noise source, which is given by:

$$\sqrt{<i^2>} = 2qI\,\Delta f(P_0 R + I_{eq})$$

where the equivalent dark current also includes the Johnson noise of the resistor, R_{eq}:

$$I_{eq} = (I_{dark} + \frac{2V_t}{R_{eq}})$$

The signal to noise ratio is then given by:

$$\frac{S}{N} = \frac{i_{ph}^2 R_{eq}}{<i^2> R_{eq}} = \frac{m^2 P_0^2 R^2}{2q\Delta f(P_0 R + I_{eq})}$$

from the above equation one can find the required optical power P_0 needed to obtain a given signal to noise ratio, S/N:

$$P_0 = (S/N)\frac{2q\Delta f}{m^2 \mathcal{R}}\{1 + \sqrt{1 + \frac{I_{eq}m^2}{q\Delta f(S/N)}}\}$$

The noise equivalent power is now defined as the ac (RMS) optical power needed to obtain a signal-to-noise ratio of one for a bandwidth of 1 Hz or:

$$NEP = \frac{mP_0}{\sqrt{2}}(\Delta f = 1, S/N = 1) = \frac{\sqrt{2q}}{m\mathcal{R}}\{1 + \sqrt{1 + \frac{I_{eq}m^2}{q}}\}$$

We now consider two limiting case in which the NEP is either limited by the optical power or by the dark current.
For $\frac{I_{eq}m^2}{q} << 1$ it is the average optical power rather than the dark current which limits the NEP:

$$NEP = \frac{2\sqrt{2q}}{m\mathcal{R}}$$

The noise equivalent power can also be used to calculate the ac optical power if the bandwidth differs from 1Hz from:

$$\frac{mP_0}{\sqrt{2}} = NEP\,\Delta f$$

where the noise equivalent power has units of W/Hz. However, the optical power is mostly limited by the dark current for which the expressions are derived below.

For $\dfrac{I_{eq}m^2}{q} \gg 1$ it is the dark current (including the Johnson noise of the resistor) which limits the NEP:

$$\text{NEP} = \frac{\sqrt{2qI_{eq}}}{R}$$

Again one can use the noise equivalent power to calculate the minimum detectable power for a given bandwidth:

$$\frac{mP_0}{\sqrt{2}} = \text{NEP}\sqrt{\Delta f}$$

where the noise equivalent power has now units of $W/\sqrt{Hz}$.

Switching of a P-i-N Photodiode

A rigorous solution for the switching time of a p-i-n photodiode starts from the continuity equations for electrons and holes:

$$\frac{\partial n}{\partial t} = \frac{1}{q}\frac{\partial J_n}{\partial x} - \frac{np-n_i^2}{n+p+2n_i}\frac{1}{\tau_0} + g_{op}(x,t)$$

$$\frac{\partial p}{\partial t} = -\frac{1}{q}\frac{\partial J_p}{\partial x} - \frac{np-n_i^2}{n+p+2n_i}\frac{1}{\tau_0} + g_{op}(x,t)$$

with,

$$J_n = q\mu_n n\mathcal{E} + qD_p\frac{\partial n}{\partial x}$$

$$J_p = q\mu_p p\mathcal{E} - qD_n\frac{\partial p}{\partial x}$$

and the electric field is obtained from Gauss's law. For a p-i-n diode with generation only at t = 0 and neglecting recombination and diffusion these equations reduce to:

$$\frac{\partial n}{\partial t} = \frac{\partial n}{\partial x}\mu_n\mathcal{E} \text{ and } \frac{\partial p}{\partial t} = -\frac{\partial p}{\partial x}\mu_p\mathcal{E}$$

Where the electric field, $\mathcal{E}$, is assumed to be a constant equal to:

$$\mathcal{E} = \frac{\phi_i - V_a}{d}$$

replacing $n(x,t)$ by $n^*(x - v_n t)$ and $p(x,t)$ by $p^*(x - v_p t)$ yields $v_n = -\mu_n \mathcal{E}$ and $v_p = \mu_p \mathcal{E}$.

The carrier distributions therefore equal those at t = 0 but displaced by a distance $\mu_n \mathcal{E} t$ for holes and $-\mu_p \mathcal{E} t$ for electrons. The total current due to the moving charge is a displacement current which is given by:

$$I_{ph}(t) = \frac{dQ}{dt} = \iiint \frac{\rho}{d} \frac{dx}{dt} dV = \frac{A}{d} \int_0^d \rho v \, dx$$

$$J_{ph}(t) = q \frac{A}{d} \mathcal{E} \int_0^d (\mu_n n + \mu_p p) dx$$

$$J_{ph}(t) = q \frac{A(\phi_i - V_a)}{d^2} [\mu_n n + \mu_p p]$$

For $t < |d/v_n|$ and $t < |d/v_p|$. For a uniform carrier generation this reduces to:

$$I_{ph}(t) = \frac{qA(\phi_i - V_a)}{d^2} [\mu_n n_0^*(d - |v_n t|) + \mu_p p_0^*(d - |v_p t|)]$$

$$I_{ph}(t) = \frac{qA(\phi_i - V_a)}{d} [\mu_n n_0^*(1 - \frac{|v_n t|}{d} + \mu_p p_0^*(d - \frac{|v_p t|}{d})]$$

In the special case where $v_n = v_p$ or $\mu_n = \mu_p$ he full width half maximum (FWHM) of the impulse response is:

$$\text{FWHM} = \frac{d}{|v_n|^2} = \frac{d^2}{2\mu_n(\phi_i - V_a)} = \frac{t_r}{2} \text{ with } t_r = \frac{d^2}{\mu_n(\phi_i - V_a)}$$

Rule of thumb to convert a pulse response to −3 dB frequency: Assuming the photodiode response to be linear, the FWHM can be related to the half-power frequency by calculating the Fourier transform. For a gaussian pulse response (which also yields a gaussian frequency response) this relation becomes:

$$f - 3dB = \frac{440\,\text{GHz}}{\text{FWHM(in ps)}}$$

Since the bandwidth depends on the transit time, which in turn depends on the depletion layer width, there is a tradeoff between the bandwidth and the quantum efficiency.

Solution in the Presence of Drift, Diffusion and Recombination

If we simplify the SHR recombination rate to n/τ and p/τ and assume a constant electric field and initial condition $n(x,0) = n_0$, the electron concentration can be obtained by solving the continuity equation, yielding:

$$n(x,t) = e^{-\alpha x} \sum_k e^{-\xi_{kn} x} B_k \sin \frac{k\pi x}{d}$$

where,

$$B_k = \frac{2n_0 k\pi}{(k\pi)^2 + (\alpha d)^2} [1 - (-1)^k e^{\alpha d}]$$

with,

$$\xi_{kn} = D_n \{\alpha^2 + (\frac{k\pi}{d})^2\} + \frac{1}{\tau} \text{ and } \alpha = \frac{E}{2V_t}$$

For this analysis we solved the continuity equation with $n(0,t) = n(L,t) = 0$ implying infinite recombination at the edges of the depletion region. The initial carrier concentration n_0 can also be related to the total energy which is absorbed in the diode at time t = 0:

$$n_0 = \frac{E_{pulse}}{E_{ph} A d}$$

and the photo current (calculated as described above) is:

$$I_{ph}(t) = \frac{qA\mu_n(\phi_i - V_a)}{d} \sum_k C_k \exp(-\xi_k t)$$

with C_k given by:

$$C_k = \frac{2n_0(k\pi)^2[1-(-1)^k e^{\alpha d}][1-(-1)^k e^{-\alpha d}]}{[(\alpha d)^2 + (k\pi)^2]^2}$$

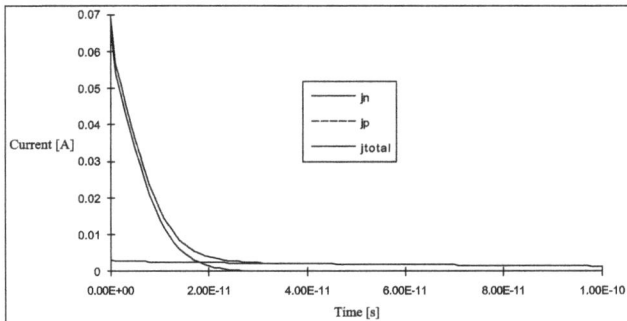

The above equations can be used to calculate the impulse response of a photodiode. Each equation must be applied to electrons as well as holes since both are generated within the diode. Typically electrons and holes have a different mobility, which results in two regions with different slopes. This effect is clearly visible in GaAs diodes as illustrated with the figure.

Figure Photocurrent calculated using equation for a GaAs diode with $\phi_i - V_a = 0.3\,\text{V}$, $E_{pulse} = 10^{-13}\,J, E_{ph} = 2\,\text{eV}$ and d = 2μm.

Harmonic Solution

Whereas above section provides a solution to the pulse response, one can also solve the frequency response when illuminating with a photon flux $\Phi_1 e^{j\omega t}$. If the photodiode has a linear response, both methods should be equivalent. To simplify the derivation, we assume that the total flux (in photons/s cm^2) is absorbed at x = 0. This is for instance the case for a p-i-n diode with a quantum well at the interface between the p-type and intrinsic region and which is illuminated with long wavelength photons, which only absorb in the quantum well. The carriers moving through the depletion region cause a conduction current $J_{cond}(x)$:

$$J_{cond}(x) = q\Phi_1 e^{j\omega(t-x/v_n)}$$

where $v_n = \mu_n \mathcal{E}$ is a constant velocity.

From Ampere's law applied to a homogeneous medium, we find:

$$\frac{1}{\mu_0}\int \vec{B}d\vec{l} = \int \vec{J}d\vec{S} + \frac{\partial}{\partial t}\int \varepsilon_s \vec{E}d\vec{S}$$

And the total current is the sum of the conduction and the displacement current:

$$J_{total} = \frac{1}{d}\int_0^d [J_{cond} + \varepsilon_s \frac{\partial E}{\partial t}]dx$$

If we assume that the electric field is independent of time, the total photo current equals:

$$J_{ph}(t) = \frac{q\Phi_1(1-e^{j\omega t_r})e^{j\omega t_r}}{j\omega t_r}$$

with transit time $t_r = \dfrac{d^2}{\mu_n(\phi_i - V_a)}$. The corresponding -3 dB frequency is:

$$f-3dB = \frac{2.78}{2\pi t_r} = \frac{443GHz}{t_r[ps]}$$

Time Response due to Carriers Generated in the Q.N. Region

For an infinitely long quasi-neutral (Q.N.) region and under stationary conditions, the generated carriers are only collected if they are generated within a diffusion length of the depletion region. The average time to diffuse over one diffusion length is the recombination time, τ. Postulating a simple exponential time response we find that the current equals:

$$I_{ph}(t) = I_{ph}(0)e^{-t/\tau}$$

Because of the relatively long carrier lifetime in fast photodiodes, carriers absorbed in the quasineutral region produce a long "tail" in the pulse response and should be avoided.

Dynamic Range of a Photodiode

The dynamic range is the ratio of the maximal optical power which can be detected to the minimal optical power. In most applications the dynamic range implies that the response is linear as well. The saturation current, defined as the maximum current which can flow through the external circuit, equals:

$$I_{sat} = \frac{\phi_i - V_a}{R}$$

which yields an optimistic upper limit for the optical power:

$$P_{max} = \frac{I_{sat}}{\mathcal{R}} = \frac{\phi_i - V_a}{\mathcal{R} R}$$

and the dynamic range is defined as the ratio of the maximum to the minimum power:

$$D.R. = \frac{P_{max}}{P_{min}}$$

Using above equation for the minimum power the dynamic range becomes independent of the responsivity and equals:

$$D.R. = \frac{\phi_i - V_a}{R\sqrt{<i^2>}}$$

For example, if the equivalent noise current equals $I_{eq} = 1$ì A, the bandwidth $\Delta f = 1\,\text{GHz}$, the impedance $R = 50\,\Omega$, and the applied voltage $V_a = 0$, then the dynamic range equals 1.35×10^6 (for $\phi_i = 1.2\,\text{V}$) or $61.3\,\text{dB}$.

Photoconductors

Photoconductors consist of a piece of semiconductor with two Ohmic contacts. Under illumination, the conductance of the semiconductor changes with the intensity of the incident optical power. The current is mainly due to majority carriers since they are free to flow across the Ohmic contacts. However the majority carrier current depends on the presence of the minority carriers. The minority carriers pile up at one of the contacts, where they cause additional injection of majority carriers until the minority carriers recombine. This effect can cause large "photoconductive" gain, which depends primarily of the ratio of the minority carrier lifetime to the majority carrier transit time. Long carrier lifetimes therefore cause large gain, but also a slow response time. The gain-bandwidth product of the photoconductor is almost independent of the minority carrier lifetime and depends only on the majority carrier transit time. Consider now a photoconductor with length, L, width W and thickness d, which is illuminated a total power, P. The optical power, P(x), in the material decreases with distance due to absorption and is described by:

$$\frac{dP(x)}{dx} = -\alpha P(x), \text{ yielding } P(x) = P(0)\exp(-\alpha x)$$

The optical power causes a generation of electrons and holes in the material. Solving the diffusion equation for the steady state case and in the absence of a current density gradient one obtains for the excess carrier densities:

$$n' = p' = \frac{\tau \alpha P(0)\exp(-\alpha x)}{h\upsilon WL}$$

Where it was assumed that the majority carriers, which primarily contribute to the photocurrent, are injected from the contacts as long as the minority carriers are present. The photo current due to the majority carriers (here assumed to be n-type) is:

$$I_n = \frac{q}{h\upsilon}(1-R)P_{in}\frac{\tau}{t_r}(1-\exp(-\alpha d))$$

where t_r is the majority carrier transit time given by:

$$t_r = \frac{L^2}{\mu V}$$

The equation above also includes the power reduction due to the reflection at the surface of the semiconductor. The normalized photocurrent is plotted in figure as a function of the normalized layer thickness for different ratio of lifetime to transit time.

As an example, consider a silicon photoconductor with $\mu_n = 1400\,cm^2/V-s$ and and $\tau = 1\mu s$. The photoconductor has a length of 10 micron and width of 100 micron.

For an applied voltage of 5 Volt, the transit time is 143 ps yielding a photoconductive gain of 7000. For a normalized distance $\alpha d = 1$ and incident power of 1 mW the photocurrent equals 1.548 mA. A reflectivity of 30% was assumed at the air/silicon interface.

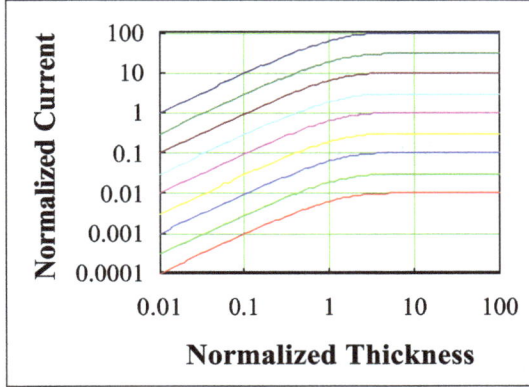

Normalized current $\dfrac{I_n}{P_{in}}\dfrac{h\upsilon}{q}$ versus normalized thickness αd as a function of the ratio of the minority carrier lifetime to the majority carrier transit time, τ / t_r, ranging from 0.01 (bottom curve) to 100 (top curve).

High photoconductive gain is typically obtained for materials with a long minority carrier lifetime, τ, high mobility, μ_n, and above all a photoconductor with a short distance, L, between the electrodes.

Metal-Semiconductor-Metal (MSM) Photodetectors

Metal-semiconductor-metal photodetectors are the simplest type of photodetectors since they can be fabricated with a single mask. They typically consist of a set of interdigitated fingers, resulting in a large active area and short distance between the electrodes.

Responsivity of a MSM Detector

The responsivity for a detector with thickness, d, surface reflectivity, R, finger spacing, L, and finger width, w, is given by:

$$\mathcal{R} = \left|\frac{I_{ph}}{P_{in}}\right| = \frac{q(1-R)L}{h\upsilon(L+w)}[1-e^{-\alpha d}]$$

Where α is the absorption length and the reflectivity, R, of the air-semiconductor interface as a function of the incident angle is given by:

$$R_{TE} = (\frac{n_1 \cos\theta_i - n_2 \cos\theta_t}{n_1 \cos\theta_i + n_2 \cos\theta})^2$$

$$R_{TM} = (\frac{n_2 \cos\theta_i - n_1 \cos\theta_t}{n_2 \cos\theta_i + n_1 \cos\theta})^2$$

with $n_2 \sin\theta_t = n_t \sin\theta_i$,

with θ_i the incident angle, and θ_t the transmitted angle. R_{TE} is the reflectivity if the electric field is parallel to the surface while R_{TM} is the reflectivity if the magnetic field is parallel to the surface. The reflectivity as a function of θ_i, for an air-GaAs interface is shown in the figure.

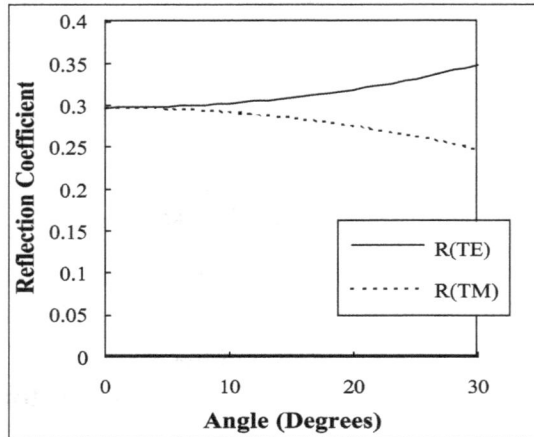

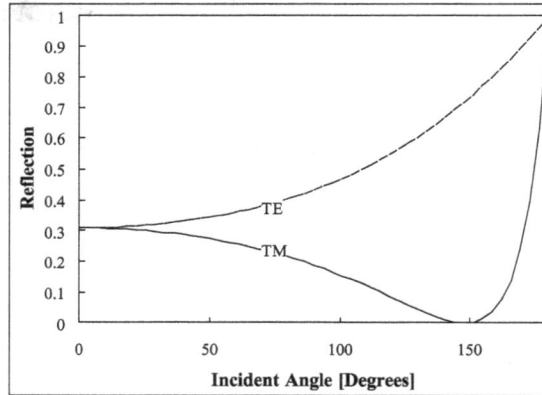

Angular dependencies of the reflectivity of an Air-to-GaAs interface Including drift, diffusion and recombination the responsivity becomes.

$$\mathcal{R} = \left|\frac{I_{ph}}{P_{in}}\right| = \frac{q(1-R)L}{h\upsilon(L+w)}[1-e^{-\alpha d}]\frac{V_a}{V_t}$$

$$\times \{[\frac{L_n^2}{L^3}\{L\frac{e^{(\beta_n-\alpha)L}}{2(\beta_n-\alpha)}(1-\coth(\beta_n L)+e^{\alpha L}\operatorname{csch}(\beta_n L)$$

$$+\frac{e^{-(\alpha+\beta_n)L}-1}{2(\alpha+\beta_n)}(1+\coth(\beta_n L)-e^{\alpha L}\operatorname{csch}(\beta_n L)\}]$$

$$[\frac{L_p^2}{L^3}\{L\frac{e^{(\beta_p-\alpha)L}-1}{2(\beta_p-\alpha)}(1-\coth(\beta_p L)+e^{\alpha L}\operatorname{csch}(\beta_p L)$$

$$+\frac{e^{-(\alpha+\beta_p)L}-1}{2(\alpha+\beta_p)}(1+\coth(\beta_p L)-e^{\alpha L}\operatorname{csch}(\beta_p L)\}]$$

with,

$$\alpha=\frac{V_a}{2V_t L}\cdot\beta_n=\sqrt{\alpha^2+\frac{1}{L_n^2}}\text{ and }\beta_p=\sqrt{\alpha^2+\frac{1}{L_p^2}}$$

The above expression can be used to calculate the current as a function of the applied voltage. An example is shown in the figure. Both the electron and the hole current are plotted as is the total current. The difference between the electron and hole current is due to the recombination of carriers. For large voltages all photo-generated carriers are swept out yielding a saturation of the photocurrent with applied voltage, whereas for small voltages around zero diffusion is found to be the dominant mechanism. The ratio of the transit time to the diffusion time determines the current around zero volt. In the absence of velocity saturation both transit times depend on the carrier mobility so that the ratio becomes independent of the carrier mobility. This causes the I-V curves to be identical for electrons and holes in the absence of recombination.

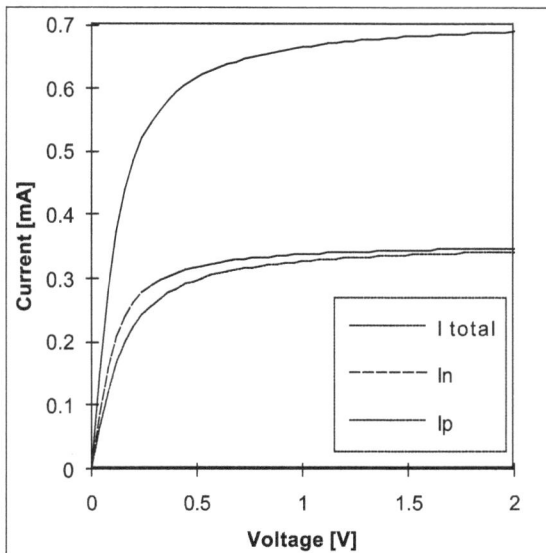

Current - Voltage characteristic of an MSM photodiode.

Pulse Response of a MSM Detector

The pulse response can be calculated by solving the time dependent continuity equation, yielding:

$$I_{ph}(t) = \frac{qAV_a}{L}\frac{(1-R)L}{L+w}[1-e^{\alpha d}]\sum_k C_k[\mu_n e^{-\xi_{kn}t} + \mu_p e^{-\xi_{kp}t}]$$

with C_k given by:

$$C_k = \frac{2n_0(k\pi)^2[1-(-1)^k e^{\alpha d}][1-(-1)^k e^{-\alpha d}]}{[(\alpha d)^2 + (k\pi)^2]^2}$$

where,

$$n_0 = \frac{E_{pulse}}{E_{ph}AL}$$

and,

$$\xi_{kn} = D_n\{\alpha^2 + (\frac{k\pi}{d})^2\} + \frac{1}{\tau}, \xi_{kp} = D_p\{\alpha^2 + (\frac{k\pi}{d})^2\} + \frac{1}{\tau} \text{ and } \alpha = \frac{\varepsilon}{2V_t}$$

This solution is plotted in the figure.

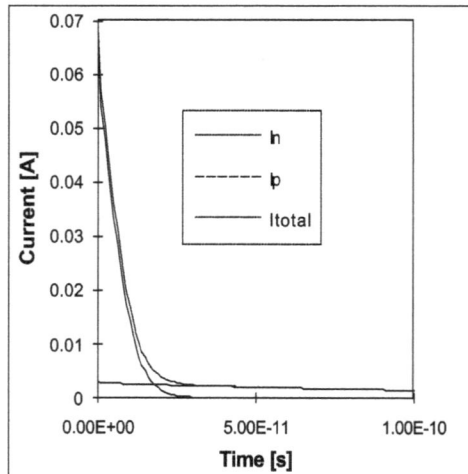

Transient behavior (pulse energy, $E_{pulse} = 0.1\text{P}^J, V_a = 0.3V$).

Equivalent Circuit of a MSM Detector

The equivalent circuit of the diode consists of the diode capacitance, C_p, a parallel resistance, R_p, obtained from the slope of the I-V characteristics at the operating voltage in parallel to the photocurrent, I_{ph}, which is obtained by calculating the convolution of the impulse response and the optical input signal. A parasitic series inductance, L_B, primarily due to the bond wire, and a series resistance, R_s, are added to complete the equivalent circuit shown in the figure.

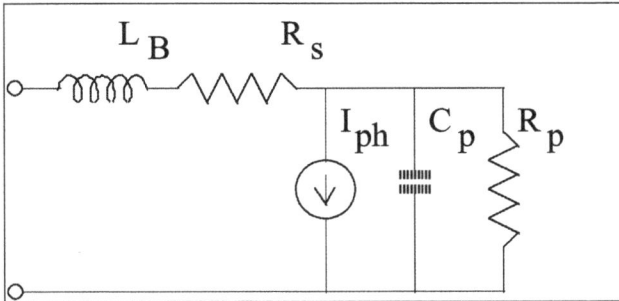

Equivalent circuit of an MSM detector.

PIN Photodiode

A PIN diode is essentially a variable resistor. To determine the value of this resistance, consider a volume comparable to a typical PIN diode chip, say 20 mil diameter and 2 mils thick. This chip has a DC resistance of about 0.75 M Ω. Note: 1 mil = 0.001 inches.

In real diodes there are impurities, typically boron, which cannot be segregated out of the crystal. Such impurities contribute carriers, holes or electrons, which are not very tightly bound to the lattice and therefore lower the resistivity of the silicon.

The resistivity of the I region and thus the diode resistance is determined by the number of free carriers within the I region. The resistivity of any semiconductor material is inversely proportional to the conductivity of the material.

Expressed mathematically the resistivity of the I region is:

$$I / P_1 = q(\mu_N N + \mu_P P)$$

where q is the electronic charge (q = 1.602 x 10^{-19} coul.), μN and μP are the mobilities of electrons and holes respectively.

Consider electrons and holes travelling in opposite directions within the I region under the impetus of an applied, positive electric field. The I region will fill up and an equilibrium condition will be reached. In non-equilibrium conditions excess minority carriers exist, and recombination between holes and electrons proceed to restore equilibrium. Recombination often occurs because of interactions between mobile charge carriers and imperfections in the semiconductor crystalline structure, either structural defects or dopant atoms. The rate of recombination of holes and electrons is proportional to the carrier concentrations and inversely proportional to a property of the semiconductor called the LIFETIME, T_L, of the minority carriers.

In the case of applied forward bias, the equation governing mobile charges in the I region is:

$$\frac{dQ_S}{dt} = I_F - \frac{Q_S}{T_L}$$

where Q_S is the stored charge and $Q_S = q(N+P)$.

Under steady conditions the mobile charge density in the I region is constant, i. e.,

$$\frac{dQ_S}{dt} = 0,$$

so that,

$$I_F = Idc = \frac{Q_S}{T_L}$$

We can next proceed to calculate the forward resistance of a PIN diode of cylindrical geometry with a thickness, W_1, and an area, A. We can ignore some details of analysis not critical to this note.

The forward current I_F was given before as:

$I_F = Q/T_L$ where Q = Charge per unit volume and

$Q = q(N+P)W_I A$ therefore,

$$I_F = \frac{q(N+P)W_I A}{T_L}$$

If there is not unneutralized charge in the I region, P= N and then:

$$I_F = \frac{2q\,NW_I A}{T_L}$$

and the resistivity of the I region, given previously, will now be:

$$P_I (2q\,\bar{\mu}N)^{-1} \text{ where } \bar{\mu} = \frac{(\mu_N + \mu_p)}{2}$$

The resistance of the I region will then be:

$$R_F = \frac{P_I W_I}{A} = (2q\,\bar{\mu}N)^{-1}\frac{W_I}{A}$$

Combining equations yields:

$$R_F = \frac{W_I^2}{2I_{F\bar{\mu}}T_L}$$

The above relation is a fundamental equation of PIN diode theory and design. Rigorous analysis such as reported by Chaffin shows:

$$R_F = \frac{2kT}{qI_F} \sinh\left(\frac{W_I}{2\sqrt{DT_L}}\right) \tan^{-1}$$

$$\left[\sinh\left(\frac{W_I}{2\sqrt{DT_L}}\right)\right]$$

Where k = Boltzmann's constant, 1. 38044 x 10^{-23} joule/kelvin.

T = Temperature in degrees kelvin

D = Diffusion constant

Typical data on R_S as a function of bias current are shown in figure. A wide range of design choices is available, as the data indicate. Many combinations of W and T_L have been developed to satisfy the full range of applications.

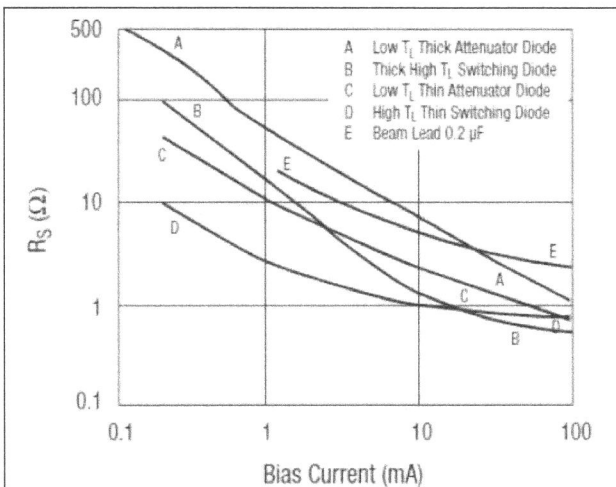

A	Low T_L Thick Attenuator Diode	
B	Thick High T_L Switching Diode	
C	Low T_L Thin Attenuator Diode	
D	High T_L Thin Switching Diode	
E	Beam Lead 0.2 µF	

Typical Series Resistance as a Function of Bias (1 GHz).

Breakdown Voltage Capacitance Q Factor

The previous section on R_S explained how a PIN can become a low resistance, or a "short". This section will describe the other state—a high impedance, or an open.

Silicon has a dielectric strength of about 400 V per mil, and all PIN diodes have a parameter called V_B, breakdown voltage, which is a direct measure of the width of the I region. Voltage in excess of this parameter results in a rapid increase in current flow (called avalanche current). When the negative bias voltage is below the breakdown of the I region, a few nanoamps will be drawn. As V_B is approached, the leakage current increases.

Typically leakage current occurs at the periphery of the I region. For this reason various passivation materials (silicon dioxide, silicon nitride, hard glass) are deposited to protect and stabilize this surface and minimize leakage. These techniques have been well advanced at Skyworks and provide a reliable PIN diode.

V_B is usually specified at a reverse current of 10 microamps.

In simplest form the capacitance of a PIN is determined by the area and width of the I region and the dielectric constant of silicon. This minimum capacitance is obtained by the application of a reverse bias in excess of V_{PT}, the voltage at which the depletion region occupies the entire I layer.

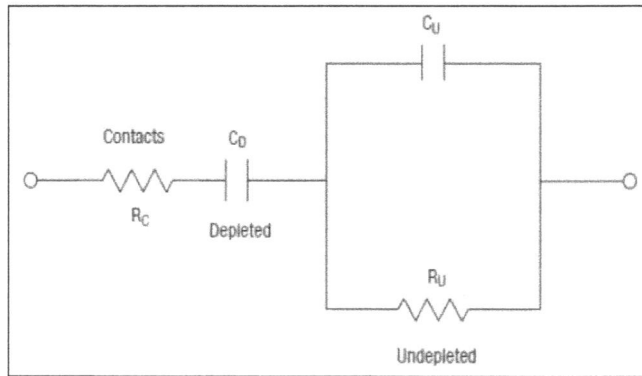

Equivalent Circuit of I Region Before Punch-Through.

Consider the undepleted region: this is a lossy dielectric consisting of a volume (area A, length L) of silicon of permittivity 12 and resistivity ρ. The capacitance is:

$\dfrac{12E_0A}{L}$, and the admittance is $\dfrac{24\pi 12E_0A}{L}$.

The resistance is proportional to L/A and the conductance to A/L. At voltages below V_{PT}, C_J will increase and approach ∞ capacitance at a forward bias of 0.7 V in silicon and 0.9 V in GaAs.

Skyworks measures junction capacitance at 1 MHz; this is a measure of the depletion zone capacitance.

For I region thickness of W and a depletion width X_d, the undepleted region is (W-X_d).

The capacitance of the depleted zone is, proportionally,

$\dfrac{1}{X_d}$, of the undepleted, $\dfrac{1}{W - X_d}$.

The 1 MHz capacitance decreases with bias until "punchthrough" where X_d = W. At microwave frequencies well above the crossover, the junction looks like two capacitors in series.

$C_T = \dfrac{C_d C_U}{C_d + C_U}$, which is proportional to 1/W i. e., the microwave capacitance tends to

be constant, independent of X_d and bias voltage.

However, since the undepleted zone is lossy, an increase in reverse bias to the punch-through voltage reduces the RF power loss.

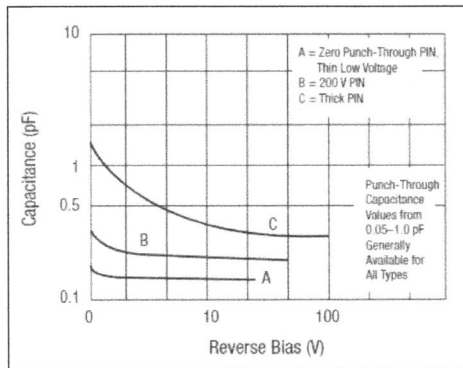

Typical Capacitance.

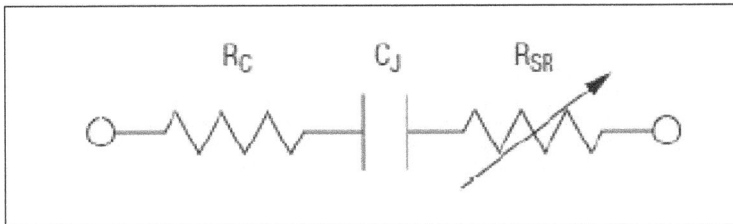

Simplified Equivalent Circuit, Series.

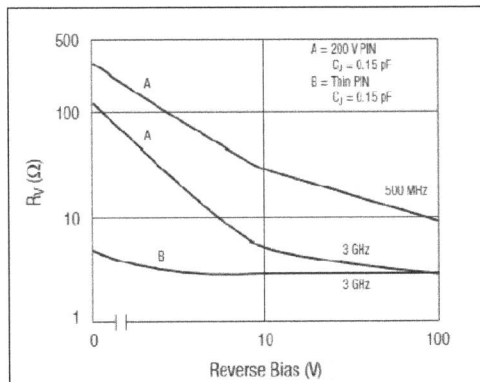

Reverse Series Resistance.

A good way to understand the effects of series resistance is to observe the insertion loss of a PIN chip series mounted in a 50 Ω line.

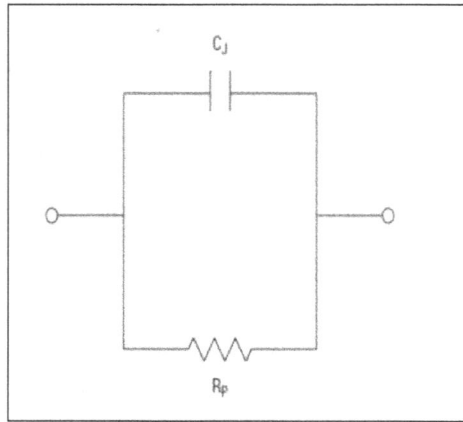

Simplified Equivalent Circuit, Shunt.

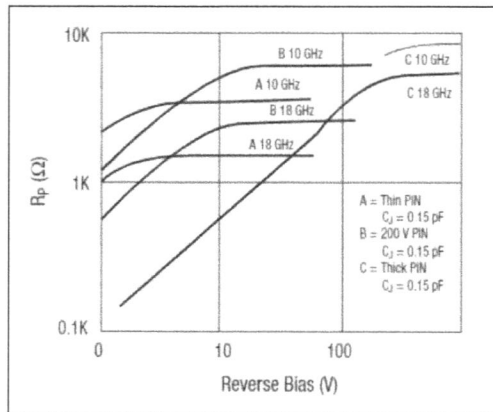

Reverse Shunt Resistance.

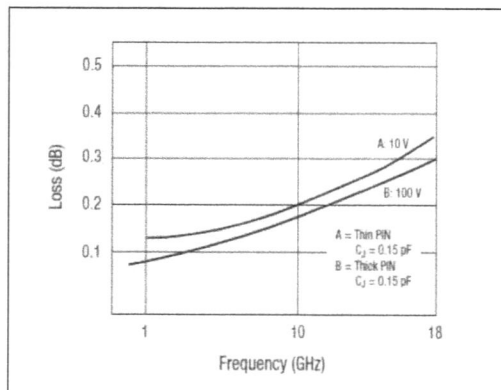

Insertion Loss vs. Frequency.

An accepted way to include reverse loss in the figure of merit of a PIN is to write the switching cut off frequency,

$$F_{CS} = \frac{1}{2\pi C_T \sqrt{R_S R_V}}$$

where R_S and R_V are measured under the expected forward and reverse bias conditions at the frequency of interest.

The punch-through voltage is a function of the resistivity and thickness of the I region. It is advisable to measure loss as a function of bias voltage and RF voltage to determine if the correct diode has been selected for your application.

Switching Considerations

Consider a PIN diode and a typical drive circuit. When the system calls for a change in state, the logic command is applied to the driver. There is delay time in the driver, in the passive components as well as in the transistors, before the voltage at Point A begins to change. There is a further delay before that voltage has stabilized. Most diode switching measurements are measured with the time reference being the 50% point of the (Point A) command waveform.

The diode begins to respond immediately, but there is a delay before the RF impedance begins to change. It is the change in impedance that causes the RF output to switch.

The driver waveforms shown are required for the fastest total switching times.

Reverse to Forward

In the high impedance state, the IV characteristics are inductive. This can be considered a function of the fact that the I region must become flooded with stored charge before the current (and RF impedance) stabilizes. Accordingly, the driver must deliver a current spike with substantial overvoltage. The capacitor paralleling the output dropping resistor is called a "speed-up" capacitor and provides the spike.

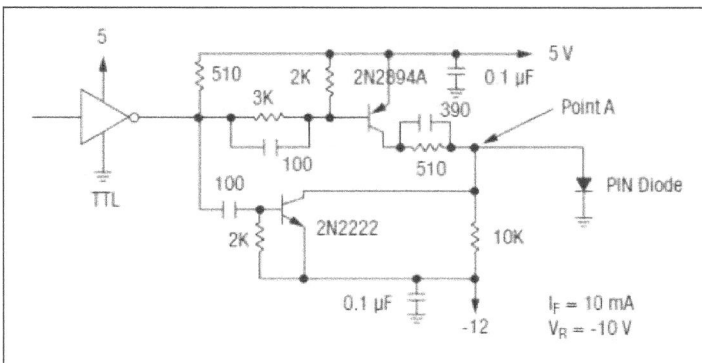

SPST Switch Driver for 10 ns.

Typical total switching time can be on the order of 2% to 10% of the specified diode lifetime and in general is much faster than switching in the other direction, from forward to reverse.

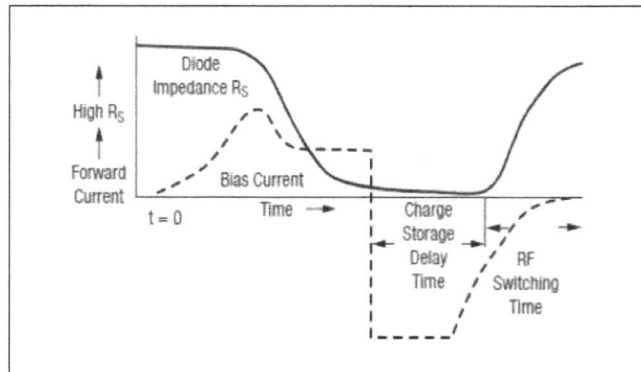

PIN Diode Switch Waveforms.

Forward to Reverse

In this mode the problem is to extract the stored charge rapidly. Once again the solution is a reverse current spike coupled with a moderately high reverse bias voltage, with reverse current on the order of 10 to 20 times forward bias,

$$I_I I_R = 0.10 \text{ to } 0.05 \text{ or less}$$

The charge storage delay will be 5% to 10% of the lifetime. Additionally the actual RF switching time will be minimized by a large negative bias and/or by a low forward bias.

Bias Circuitry

It is advisable to design the bias circuit to have the same characteristic impedance as the RF line to minimize reflections and ringing. Extraneous capacitance, in the form of blocking and bypass elements, must not be excessive. A typical 60+ pF bypass in a 50 Ω RF circuit produces a 3.0 nanosecond rise time. A few of these make it impossible to exploit the fastest PINs.

Temperature Effects on Forward Resistance

Series Resistance

Two conflicting mechanisms influence temperature behavior. First, as temperature rises, lifetime increases, allowing a greater carrier concentration and lowering R_s. Secondly, however, at higher temperature change, mobility decreases, raising R_s. The net result of these competing phenomena is a function of diode design, bias current, RF power level, and frequency.

The figure shows unlabeled curves of R_s vs. temperature with bias as a parameter. Most diodes show a monotonic increase of series resistance as temperature increases, while reverse losses tend to increase.

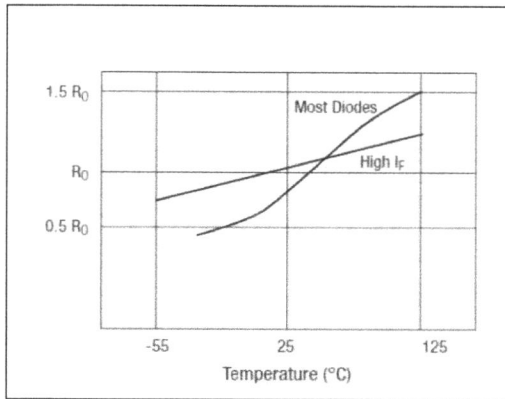

Series Resistance vs. Temperature.

A typical fast switching diode will draw 10 mA at 850 mV at 25 °C. At -55 °C, the same V_F will draw about 500 microamps; at 100 °C, I_F will be 200 mA.

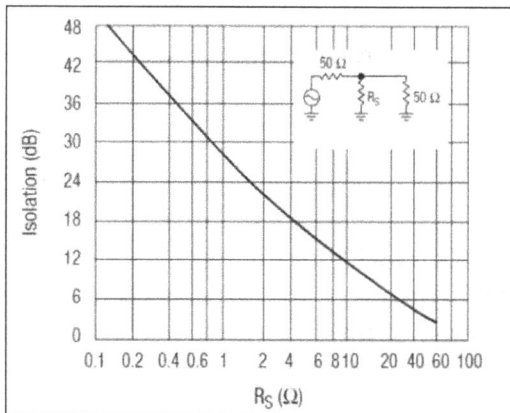

Shunt Diode Isolation vs. Forward Biased Resistance.

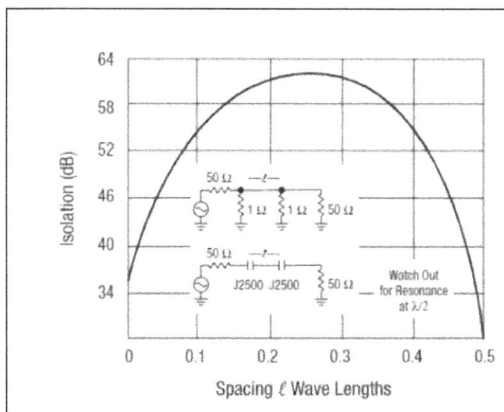

Isolation vs. Diode Spacing.

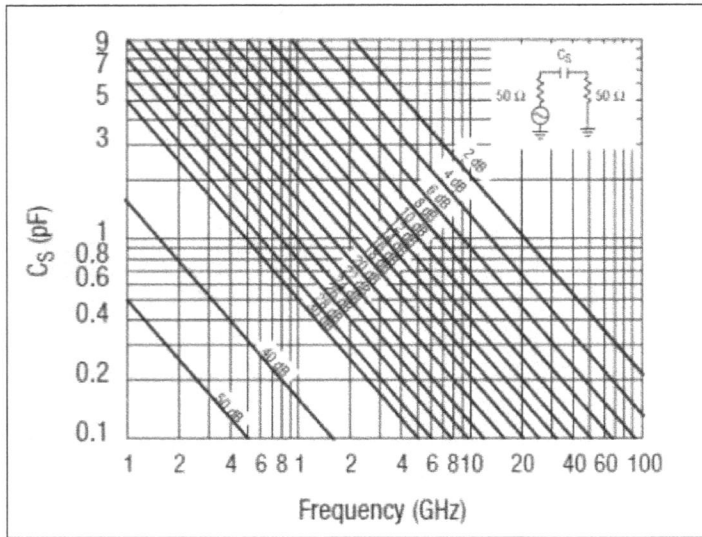

Isolation vs. Series Capacitance.

How to Specify PIN Diodes

The uses for PIN diodes fall into three categories: series element, shunt element, and a limiter diode. The following guidelines should help in specifying a PIN diode for these applications.

For Series Diodes

R_S—The forward series resistance will determine the minimum loss in the insertion loss state. Normally the diode will have a resistance slightly higher than R_S due to the internal junction resistance because of limited forward current. The ideal PIN diode series resistance is low, however. Low series resistances are associated with high idle state capacitance and a trade-off must, therefore, be made between off-state capacitance (C_J) and series resistance (R_S). Skyworks does this by choosing the proper junction diameter for your application.

C_J—The capacitance (specified at 1 MHz at punch-through) is the off state capacitance, and for a series element determines the broadband isolation or, for narrow-band applications, the bandwidth of the switch.

Shunt Elements

R_S—The forward series resistance of the shunt element determines the maximum isolation that can be obtained from this element. The ideal diode has extremely low R_S; however, diodes with low R_S have an associated capacitance (C_J) which may be high. The shunt element trade-off is to balance the required isolation with the effective insertion loss of a broadband switch at the band width of a narrow-band switch.

C_J—The capacitance (specified at 1 MHz at punch-through) needs to be a low value to maintain low loss, broadband switching. C_J will also determine the input VSWR of the switch for broadband applications.

V_B—The breakdown voltage of a PIN diode must be specified to assure the power handling of the switch component. In general the voltage must be high enough to prevent breakdown during the reverse bias condition, including the DC applied bias and the peak RF voltage. Failure to do so will cause a condition that can result in diode limiting and under severe circumstances can cause failure. For a simple shunt switch, a 100 V breakdown diode biased at 50 V can accommodate a peak voltage of ~50 V or power of 25 W average in a 50 Ω system. For maximum power handling, the reverse bias should be one-half of V_B.

Limiters

PIN diodes with low breakdown voltages can serve as power limiters. The onset of limiting is primarily determined by the V_B of the diode. The limiting function is also affected by the lifetime of the base region. For this reason the frequency range, peak power requirements and threshold need to be specified to choose a diode with minimum leakage. The power handling capability of limiter diodes is determined by V_B for short pulse applications and thermal heat sinking for long pulsed applications. The heat sinking is a composite of the thermal path in the diode and the mounting thermal resistance. Beam-lead diodes provide very low power handling capability due to the extremely high (1200 °C/W) thermal resistance—high powers can be achieved from shunt chips which can have thermal resistances below 5 °C/W. To determine the power handling capability of a limiter circuit one needs to determine the maximum power absorbed by the limiter diode. In hard limiting, the diode will reflect most of the power and only a small portion will be absorbed.

Avalanche Photodiode

Avalanche photodiode detectors have and will continue to be used in many diverse applications such as laser range finders and photon correlation studies.

For low-light detection in the 200 to 1150 nm range, the designer has three basic detector choices - the silicon PIN detector, the silicon avalanche photodiode (APD) and the photomultiplier tube (PMT). APDs are widely used in instrumentation and aerospace applications, offering a combination of high speed and high sensitivity unmatched by PIN detectors, and quantum efficiencies at > 400 nm unmatched by PMTs.

Recently demonstrated performance includes:

- Noise equivalent power (NEP) of $<10^{-15} W/Hz^{1/2}$ for a 0.5 mm APD.

- Detection of 100-photon, 20 ns pulses with standard APDs.

- Detection of 10-photon, 20 ns pulses with special APDs.

- BER data communications at 810 nm at 60 Mb/s with only 39 photons/bit.

- Photon-counting detection efficiencies > 70% at 633 nm; dark counts of only 1.

- Count/s on 150 μm diameter APDs.

- Detection and resolution of low energy, 1 to 30 keV X-rays.

APD Structures

In order to understand why more than one APD structure exists, it is important to appreciate the design trade-offs that must be accommodated by the APD designer. The ideal APD would have zero dark noise, no excess noise, broad spectral and frequency response, a gain range from 1 to 106 or more, and low cost. More simply, an ideal APD would be a good PIN photodiode with gain. In reality however, this is difficult to achieve because of the need to trade-off conflicting design requirements. What some of these tradeoffs are, and how they are optimized in commercially available APDs, are listed below. Consider the schematic cross-section for a typical APD structure shown in figure. The basic structural elements provided by the APD designer include an absorption region A, and a multiplication region M. Present across region A is an electric field E that serves to separate the photo-generated holes and electrons, and sweeps one carrier towards the multiplication region. The multiplication region M is designed to exhibit a high electric field so as to provide internal photo-current gain by impact ionization. This gain region must be broad enough to provide a useful gain, M, of at least 100 for silicon APDs, or 10-40 for germanium or InGaAs APDs. In addition, the multiplying electric field profile must enable effective gain to be achieved at at field strength below the breakdown field of the diode.

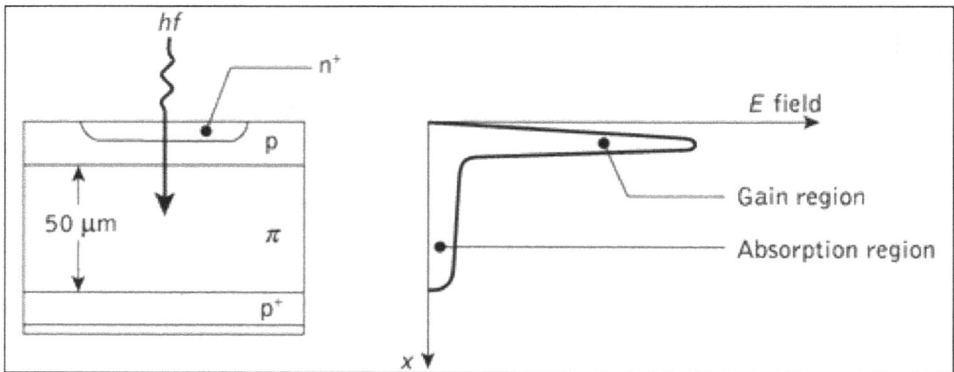

Reach-through APD Structure.

Figure shows the reach-through structure patented by PerkinElmer which offers the best available combination of high speed, low noise and capacitance and extended red response.

Critical Performance Parameters

An APD differs from a PIN photodiode by providing internal photo-electronic signal gain. Therefore, output signal current, I_S, from and APD equals $I_S = MR_0(l)P_S$, where $R_0(l)$ is the intrinsic responsivity of the APD at a gain M=1 and wavelength l, M is the gain of the APD, and P_S is the incident optical power. The gain is a function of the APDs reverse voltage, V_R, and will vary with applied bias. A typical gain-voltage curve for a silicon APD manufactured by PerkinElmer is shown in figure.

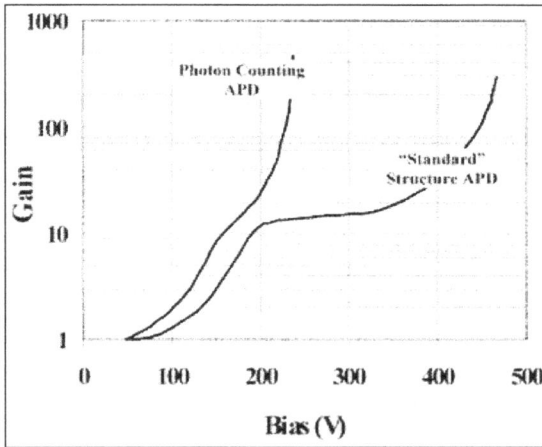

Typical gain-voltage curve for Si APDs.

One of the key parameters to consider when selecting an APD is the detector's spectral noise. Like other detectors, and APD will normally be operating in one of two noise-limited detection regimes; either detector noise limited at low power levels, or photon shot noise limited at higher powers. As an APD is designed to be operated under a reverse bias, sensitivity at low light levels will be limited by the shot noise and the APDs leakage current. Shot noise derives from the random statistical Poissonian fluctuations of the dark current, I_D (or signal current). Dark current or shot noise $I_{N(SHOT)}$ is normally given by $I_{N(SHOT)} = (2qBI_D)^{1/2}$ for a PIN detector, where B is the system bandwidth. This differs for an APD however, as bulk leakage current I_{DB}, is multiplied by the gain, M, of the APD[4]. Total leakage current I_D is therefore equal to:

$$I_D = I_{DS} + I_{DB}M$$

Where I_{DS} is the surface leakage current.

In addition, the avalanche process statistics generate current fluctuations, and APD performance is degraded by an excess noise factor (F) compared to a PIN. A detailed explanation and calculation of the excess noise factor, F. The total spectral noise current for an APD in dark conditions is thus given by:

$$I_N = [2q(I_{DS} + I_{DB}M^2F)B]^{1/2}$$

Where q is the electron charge.

At higher signal light levels, the detector transitions to the photon shot noise limited regime where sensitivity is limited by photon shot noise on the current generated by the optical signal. Total noise from the APD in illuminated conditions will therefore equal the quadratic sum of the detector noise plus the signal shot noise. For a given signal power, P_S, this is given by:

$$I_{N(TOTAL)} = [2q(I_{DS} + (I_{DB}M^2 + R_0(I)M^2P_S)F)B]^{1/2}$$

In the absence of other noise sources, an APD therefore provides a signal-to-noise ratio (SNR) which is $F^{1/2}$ worse than a PIN detector with the same quantum efficiency. An APD, however, can produce a better overall system signal-to-noise ratio than a PIN detector in cases where the APD internal gain boosts the signal level without dramatically affecting the overall system noise.

Noise equivalent power (NEP) cannot be used as the only measure of a detector's relative performance, but rather detector signal-to-noise (SNR) at a specific wavelength and bandwidth should be used to determine the optimum detector type for a given application. Note that optimum signal-to-noise occurs at a gain M where total detector noise equals the input noise of the amplifier or load resistor. The optimum gain depends in part on the excess noise factor, F, of the APD, and ranges from M=100 to 1000 for silicon APDs and is limited to M=30 to 40 for germanium and InGaAs APDs.

Selecting an APD

Specifying your Requirement

APDs are general y recommended for high bandwidth applications or where internal gain is needed to overcome high pre-amp noise. The following is a simple guide that can be used to decide whether an APD is the most appropriate for one's light detection requirements.

- Determine the wavelength range to be covered.

- Determine the minimum size of the detector that can be used in the optical system. Effective optics can often be more cost-effective than the use of an overly large PIN or Avalanche photodiode.

- Determine the required electrical frequency bandwidth of the system; again, overspecifying bandwidth will degrade the SNR of the system.

Types of APDs

Avalanche photodiodes are commercially available that span the wavelength range from 300 to 1700 nm. Silicon APDs can be used between 300 to 1100 nm, germanium between 800 and 1600 nm and InGaAs from 900 to 1700 nm.

Although significantly more expensive than germanium APDs, InGaAs APDs are typically available with significantly lower noise current, exhibit extended spectral response to 1700 nm, and provide higher frequency bandwidth for a given active area. A germanium APD is recommended for environmental applications in high electro-magnetic interference (EMI), where amplifier noise is significantly higher than the noise from an InGaAs APD, or for applications where cost is primordial consideration.

Understanding the Specifications

It is important to understand several performance metrics when selecting an APD appropriate for one's application. Listed below are some of the most important specifications, and an explanation as to how different manufacturer's listings should be compared.

Responsivity and Gain: APD gain will vary as a function of applied reverse voltage, as shown in figure. In addition, for many APDs, it is not possible, or practical, to make an accurate measurement of the intrinsic responsivity, $R_o(l)$, at a gain M=1. It is therefore inappropriate to state typical gain and diode sensitivity at M=1 as a method for specifying diode responsivity at a given operating voltage. In order to characterize APD response, one must specify APD responsivity (in Amps/Watt) at a given operating voltage. However, because of diode to diode variations in the exact gain-voltage curve of each APD, the specific operating voltage for a given responsivity will vary from one APD to another. Manufacturers should therefore specify a voltage range within which a specific responsivity will be achieved. An example of a typically correct specifications for diode responsivity, in this case for an InGaAs APD, is as follows:

$$R_{MIN}(1300\,nm) = 9.0\,A/W, V_{OP} = 50\,V\,to\,90\,V, M \sim 10$$

Dark Current and Noise Current: As can be seen in the equation above, total APD dark current (and corresponding spectral noise current) is only meaningful when specified at a given operating gain. Dark current M=1 is dominated by surface current, and may be significantly less than I_{DB} x M. Since APD dark and spectral noise current are a strong function of APD gain, these should be specified at a stated responsivity level. An example of a typically correct specification for diode dark current and noise current, in this case for an InGaAs APD is as follows:

$$I_D(R = 9.0\,A/W) = 10\,nA(max), M \sim 10$$

$$I_N(R = 6.0\,A/W), 1\,MHz, 1\,Hz\,BW) = 0.8\,pA/Hz^{1/2}(max), M > 5$$

Excess Noise Factor

All avalanche photodiodes generate excess noise due to the statistical nature of the avalanche process. The Excess Noise Factor is generally denoted as F. As shown in the noise equation, $F^{\frac{1}{2}}$ is the factor by which the statistical noise on the APD current (equal to the sum of the multiplied photocurrent plus the multiplied APD bulk dark current) exceeds that which would be expected from a noiseless multiplier on the basis of Poissonian statistics (shot noise) alone.

The excess noise factor is a function of the carrier ionization ratio, k, where k is usually defined as the ratio of hole to electron ionization probabilities (k < 1). The excess noise factor may be calculated using the model developed my McIntyre which considers the statistical nature of avalanche multiplication. The excess noise factor is given by:

$$ F = k_{EFF} M + \left(1 - k_{EFF}\right)\left(2 - \frac{1}{M}\right) $$

Therefore, the lower the values of k and M, the lower the excess noise factor. The effective k-factor, k_{EFF}, for an APD can be measured experimentally by fitting the McIntyre formula to the measured dependence of the excess noise factor on gain. This is best done under illuminated conditions. It may also be theoretically calculated from the carrier ionization coefficients and the electric field profile of the APD structure.

The ionization ratio k is a strong function of the electric field across the APD structure, and takes its lowest value at low electric fields (only in Silicon). Since the electric field profile depends upon the doping profile, the k factor is also a function of the doping profile. Depending on the APD structure, the electric field profile traversed by a photo-generated carrier and subsequent avalanche-ionized carriers may therefore vary according to photon absorption depth. For indirect bandgap semiconductors such as silicon, the absorption coefficient varies slowly at the longer wavelengths, and the mean absorption depth is therefore a function of wavelength. The value of k_{EFF}, and gain, M, for a silicon APD is thus a function of wavelength for some doping profiles.

The McIntyre formula can be approximated for a k < 0.1 and M > 20 without significant loss of accuracy as:

F = 2 + kM

Also often quoted by APD manufacturers is an empirical formula used to calculate the excess noise factor, given as:

F = MX

where the value of X is derived as a log-normal linear fit of measured F-values for given values of gain M. This approximation is sufficiently good for many applications, particularly when used with APDs with a high k factor, such as InGaAs and Germanium APDs. Table provides typical values of k, X and F for silicon, germanium and InGaAs APDs. With the exception of Super-low k (SLiK™) APDs used in photon counting modules, all PerkinElmer silicon APDs have k = 0.22. Note that for germanium and InGaAs, a k-value is generally quoted at M = 10, which somewhat overestimates F at M 10.

Table: Typical values of k, X and F for Si, Ge and InGaAs APDs.

Detector Type	Ionization Ratio (k)	X-Factor	Typical Gain (M)	Excess Noise Factor (F)
Silicon (reach through structure)	0.02	-	150	4.9
Silicon (reach through structure)	0.002	-	500	3.0
Germanium	0.9	0.95	10	9.2
InGaAs	0.45	0.7 - 0.75	10	5.5

Geiger Mode

In the Geiger mode, an APD is biased above its breakdown voltage ($V_R > V_{BR}$) for operation at very high gain (typically 10^5 to 10^6). When biased above breakdown, an APD will normally conduct a large current. However, if this current is limited to less than the APDs latching current, there is a strong statistical probability that the current will fluctuate to zero in the multiplication region, and the APD will then remain in the off state until an avalanche pulse is triggered by either a bulk or photo-generated carrier. If the number of bulk carrier generated pulses is low, the APD will then remain in the off state until an avalanche pulse is triggered by either a bulk or photo-generated carrier. If the number of bulk carrier generated pulses is low, the APD can therefore be used to count individual current pulses from incident photons. The value of the bulk dark current is therefore a significant parameter in selecting an APD for photon-counting, and can be reduced exponentially by cooling. To date, only silicon APDs have bulk dark currents low enough to be suitable for use in commercially available single photon-counting modules (SPCMs), although photon-counting has been demonstrated experimentally with germanium and InGaAs APDs. Dark counts as low as < 1 cps (counts per second) have been reported for PerkinElmer SLiK™ silicon APDs cooled to -20°C, and counts of < 250 cps are typically achievable with a cooled 0.5 mm C30902S APD.

Applications

In both modes of APD operation, ie. Linear and Geiger, APDs have and will continue to be used in many diverse applications. In the linear mode operation, the APD is well suited for applications which require high sensitivity and fast response times. For example, laser range finders which incorporate APD detectors result in more sensitive instruments than ones which use conventional PIN detectors. In addition, APDs used in the application can operate with lower light levels and shorter laser pulses, thus making the range finder more eye safe.

Other applications for APDs include fast receiver modules, confocal microscopy and particle detection. A silicon APD can be used in applications to detect alpha particles, electrons with energies as high as 150 keV, and other forms of radiation.

Silicon APDs operated in the Geiger mode are used to detect single photons for photon correlation studies and are capable of achieving very short (20-50 ps) resolving times. Operated in this mode, EG&G's SLiK™ detector provide gains of up to 10^8 and quantum efficiencies of approximately 70% at 633 nm and 50% at 830 nm. Other applications in which APDs operated in this mode are used include: Lidar, Astronomical observation, Optical range finding, Optical fiber test and fault location, Ultrasensitive fluorescence, etc.

Schottky Diode

The Schottky diode (named after the German physicist Walter H. Schottky), also known as Schottky barrier diode or hot-carrier diode, is a semiconductor diode formed by the junction of a semiconductor with a metal. It has a low forward voltage drop and a very fast switching action. The cat's-whisker detectors used in the early days of wireless and metal rectifiers used in early power applications can be considered primitive Schottky diodes.

When sufficient forward voltage is applied, a current flows in the forward direction. A silicon diode has a typical forward voltage of 600–700 mV, while the Schottky's forward voltage is 150–450 mV. This lower forward voltage requirement allows higher switching speeds and better system efficiency.

Construction

1N5822 Schottky diode with cut-open packaging. The semiconductor in the center makes a Schottky barrier against one metal electrode (providing rectifying action) and an ohmic contact with the other electrode.

A metal–semiconductor junction is formed between a metal and a semiconductor, creating a Schottky barrier (instead of a semiconductor–semiconductor junction as in conventional diodes). Typical metals used are molybdenum, platinum, chromium or tungsten, and certain silicides (e.g., palladium silicide and platinum silicide), whereas the semiconductor would typically be n-type silicon. The metal side acts as the anode, and n-type semiconductor acts as the cathode of the diode; meaning conventional current can flow from the metal side to the semiconductor side, but not in the opposite direction. This Schottky barrier results in both very fast switching and low forward voltage drop.

The choice of the combination of the metal and semiconductor determines the forward voltage of the diode. Both n- and p-type semiconductors can develop Schottky barriers. However, the p-type typically has a much lower forward voltage. As the reverse leakage current increases dramatically with lowering the forward voltage, it cannot be too low, so the usually employed range is about 0.5–0.7 V, and p-type semiconductors are employed only rarely. Titanium silicide and other refractory silicides, which are able to withstand the temperatures needed for source/drain annealing in CMOS processes, usually have too low a forward voltage to be useful, so processes using these silicides therefore usually do not offer Schottky diodes.

With increased doping of the semiconductor, the width of the depletion region drops. Below a certain width, the charge carriers can tunnel through the depletion region. At very high doping levels, the junction does not behave as a rectifier any more and becomes an ohmic contact. This can be used for the simultaneous formation of ohmic contacts and diodes, as a diode will form between the silicide and lightly doped n-type region, and an ohmic contact will form between the silicide and the heavily doped n- or p-type region. Lightly doped p-type regions pose a problem, as the resulting contact has too high a resistance for a good ohmic contact, but too low a forward voltage and too high a reverse leakage to make a good diode.

As the edges of the Schottky contact are fairly sharp, a high electric field gradient occurs around them, which limits how large the reverse breakdown voltage threshold can be. Various strategies are used, from guard rings to overlaps of metallization to spread out the field gradient. The guard rings consume valuable die area and are used primarily for larger higher-voltage diodes, while overlapping metallization is employed primarily with smaller low-voltage diodes.

Schottky diodes are often used as antisaturation clamps in Schottky transistors. Schottky diodes made from palladium silicide (PdSi) are excellent due to their lower forward voltage (which has to be lower than the forward voltage of the base-collector junction). The Schottky temperature coefficient is lower than the coefficient of the B–C junction, which limits the use of PdSi at higher temperatures.

For power Schottky diodes, the parasitic resistances of the buried n+ layer and the epitaxial n-type layer become important. The resistance of the epitaxial layer is more important than it is for a transistor, as the current must cross its entire thickness. However, it serves as a distributed ballasting resistor over the entire area of the junction and, under usual conditions, prevents localized thermal runaway.

In comparison with the power p–n diodes the Schottky diodes are less rugged. The junction is direct contact with the thermally sensitive metallization, a Schottky diode can therefore dissipate less power than an equivalent-size p-n counterpart with a deep-buried junction before failing (especially during reverse breakdown). The relative advantage of the lower forward voltage of Schottky diodes is diminished at higher forward currents, where the voltage drop is dominated by the series resistance.

Reverse Recovery Time

The most important difference between the p-n diode and the Schottky diode is the reverse recovery time (t_{rr}), when the diode switches from the conducting to the non-conducting state. In a p–n diode, the reverse recovery time can be in the order of several microseconds to less than 100 ns for fast diodes. Schottky diodes do not have a recovery time, as there is nothing to recover from (i.e., there is no charge carrier depletion region at the junction). The switching time is ~100 ps for the small-signal diodes, and up to tens of nanoseconds for special high-capacity power diodes. With p–n-junction switching, there is also a reverse recovery current, which in high-power semiconductors brings increased EMI noise. With Schottky diodes, switching is essentially "instantaneous" with only a slight capacitive loading, which is much less of a concern.

This "instantaneous" switching is not always the case. In higher voltage Schottky devices, in particular, the guard ring structure needed to control breakdown field geometry creates a parasitic p-n diode with the usual recovery time attributes. As long as this guard ring diode is not forward biased, it adds only capacitance. If the Schottky junction is driven hard enough however, the forward voltage eventually will bias both diodes forward and actual t_{rr} will be greatly impacted.

It is often said that the Schottky diode is a "majority carrier" semiconductor device. This means that if the semiconductor body is a doped n-type, only the n-type carriers (mobile electrons) play a significant role in normal operation of the device. The majority carriers are quickly injected into the conduction band of the metal contact on the other side of the diode to become free moving electrons. Therefore, no slow random recombination of n and p type carriers is involved, so that this diode can cease conduction faster than an ordinary p–n rectifier diode. This property in turn allows a smaller device area, which also makes for a faster transition. This is another reason why Schottky diodes are useful in switch-mode power converters: the high speed of the diode means that the circuit can operate at frequencies in the range 200 kHz to 2 MHz, allowing the use of small inductors and capacitors with greater efficiency than would be possible with other diode types. Small-area Schottky diodes are the heart of RF detectors and mixers, which often operate at frequencies up to 50 GHz.

Limitations

The most evident limitations of Schottky diodes are their relatively low reverse voltage ratings, and their relatively high reverse leakage current. For silicon-metal Schottky diodes, the reverse voltage is typically 50 V or less. Some higher-voltage designs are available (200 V is considered a high reverse voltage). Reverse leakage current, since it increases with temperature, leads to a thermal instability issue. This often limits the useful reverse voltage to well below the actual rating.

While higher reverse voltages are achievable, they would present a higher forward voltage, comparable to other types of standard diodes. Such Schottky diodes would have no advantage unless great switching speed is required.

Silicon Carbide Schottky Diode

Schottky diodes constructed from silicon carbide have a much lower reverse leakage current than silicon Schottky diodes, as well as higher forward voltage (about 1.4-1.8V at 25 °C) and reverse voltage. As of 2011 they were available from manufacturers in variants up to 1700 V of reverse voltage.

Silicon carbide has a high thermal conductivity, and temperature has little influence on its switching and thermal characteristics. With special packaging, silicon carbide Schottky diodes can operate at junction temperatures of over 500 K (about 200 °C), which allows passive radiative cooling in aerospace applications.

Applications

Voltage Clamping

While standard silicon diodes have a forward voltage drop of about 0.6 V and germanium diodes 0.2 V, Schottky diodes' voltage drop at forward biases of around 1 mA is in the range of 0.15 V to 0.46 V, which makes them useful in voltage clamping applications and prevention of transistor saturation. This is due to the higher current density in the Schottky diode.

Reverse Current and Discharge Protection

Because of a Schottky diode's low forward voltage drop; less energy is wasted as heat, making them the most efficient choice for applications sensitive to efficiency. For instance, they are used in stand-alone ("off-grid") photovoltaic (PV) systems to prevent batteries from discharging through the solar panels at night, called "blocking diodes". They are also used in grid-connected systems with multiple strings connected in parallel, in order to prevent reverse current flowing from adjacent strings through shaded strings if the "bypass diodes" have failed.

Switched-mode Power Supplies

Schottky diodes are also used as rectifiers in switched-mode power supplies. The low forward voltage and fast recovery time leads to increased efficiency.

They can also be used in power supply "OR"ing circuits in products that have both an internal battery and a mains adapter input, or similar. However, the high reverse leakage current presents a problem in this case, as any high-impedance voltage sensing circuit (e.g., monitoring the battery voltage or detecting whether a mains adapter is present) will see the voltage from the other power source through the diode leakage.

Sample-and-Hold Circuits

Schottky diodes can be used in diode-bridge based sample and hold circuits. When compared to regular p-n junction based diode bridges, Schottky diodes can offer advantages. A forward-biased Schottky diode does not have any minority carrier charge storage. This allows them to switch more quickly than regular diodes, resulting in lower transition time from the sample to the hold step. The absence of minority carrier charge storage also results in a lower hold step or sampling error, resulting in a more accurate sample at the output.

Charge Control

Due to its efficient electric field control Schottky diodes can be used to accurately load or unload single electrons in semiconductor nanostructures such as quantum wells or quantum dots.

Designation

SS14 schottky diode in a DO-214AC (SMA) package (surface mount version of 1N5819).

Commonly encountered schottky diodes include the 1N58xx series rectifiers, such as the 1N581x (1 ampere) and 1N582x (3 ampere) through-hole parts, and the SS1x (1 ampere) and SS3x (3 ampere) surface-mount parts. Schottky rectifiers are available in numerous surface-mount package styles.

Small-signal schottky diodes such as the 1N5711, 1N6263, 1SS106, 1SS108, and the BAT41−43, 45−49 seriesare widely used in high-frequency applications as detectors, mixers and nonlinear elements, and have superseded germanium diodes. They are also suitable for electrostatic discharge (ESD) protection of sensitive devices such as III-V-semiconductor devices, laser diodes and, to a lesser extent, exposed lines of CMOS circuitry.

Schottky metal–semiconductor junctions are featured in the successors to the 7400 TTL family of logic devices, the 74S, 74LS and 74ALS series, where they are employed as Baker clamps in parallel with the collector-base junctions of the bipolar transistors to prevent their saturation, thereby greatly reducing their turn-off delays.

Alternatives

When less power dissipation is desired, a MOSFET and a control circuit can be used instead, in an operation mode known as active rectification.

A super diode consisting of a pn-diode or Schottky diode and an operational amplifier provides an almost perfect diode characteristic due to the effect of negative feedback, although its use is restricted to frequencies the operational amplifier used can handle.

Electrowetting

Electrowetting can be observed when a Schottky diode is formed using a droplet of liquid metal, e.g. mercury, in contact with a semiconductor, e.g. silicon. Depending on the doping type and density in the semiconductor, the droplet spreading depends on the magnitude and sign of the voltage applied to the mercury droplet. This effect has been termed 'Schottky electrowetting'.

APPLICATIONS OF PHOTODIODE

- In a simple day to day applications, photodiodes are used. The reason for their use is their linear response of photodiode to a light illumination. When more amount of light falls on the sensor, it produces high amount of current. The increase in current will be displayed on a galvanometer connected to the circuit.

- Photodiodes helps to provide an electric isolation with help of optocouplers. When two isolated circuits are illuminated by light, optocouplers is used to couple the circuit optically. But the circuits will be isolated electrically. Compared to conventional devices, optocouplers are fast.

- Photodiodes are applied in safety electronics like fire and smoke detectors. It is also used in TV units.

- When utilized in cameras, they act as photo sensors. It is used in scintillators charge-coupled devices, photoconductors, and photomultiplier tubes.

- Photodiodes are also widely used in numerous medical applications like instruments to analyze samples, detectors for computed tomography and also used in blood gas monitors.

ADVANTAGES AND DISADVANTAGES

Advantages of Photodiode

- The photodiode is linear.
- Low resistance.
- A very good spectral response.
- Better frequency response.
- Low dark current.
- Fastest photodetector.
- Long lifetime.
- Low noise.
- High quantum efficiency.
- It is highly sensitive to the light.
- Compact and lightweight.
- No high voltage requires.
- Ruggedized the mechanical stress.
- Using photodiode the speed of operation is very high.

Disadvantages of Photodiode

- Require increases in the dark current.
- It depends on the temperature.
- Small active area.
- Normal PN junction photodiode has a very high response time.
- It has very low sensitivity.
- Light sensitive device.
- Poor temperature stability.
- Change in current is very small, hence may not be sufficient to drive the circuit.
- It need offset voltage.

References

- Photodiodesymboltypes, electronic-devices-and-circuits-semiconductor-diodes: physics-and-radio-electronics.com, Retrieved 17 August, 2019

- Photodiode: electronicscoach.com, Retrieved 29 June, 2019

- Applications-of-Photodiode, photodiode-working-characteristics-applications: electronicshub.org, Retrieved 06 March, 2019

- O. D. D. Couto Jr., J. Puebla, E. A. Chekhovich, I. J. Luxmoore, C. J. Elliott, N. Babazadeh, M. S. Skolnick, and A. I. Tartakovskii Charge control in InP/(Ga,In)P single quantum dots embedded in Schottky diodes (2011), Physical Review B 84, 125301. doi:10.1103/PhysRevB.84.125301

- Advantages-and-disadvantages-of: ecstuff4u.com, Retrieved 18 February, 2019

4

Photovoltaic Cells

A specially designed electrical device that consists of semiconductor diode and converts light energy into direct current is termed as a photovoltaic cell. Its three major types are monocrystalline silicon cells, polycrystalline silicon cells and thin film cells. This chapter closely examines these types of photovoltaic cells and its related aspects to provide an extensive understanding of the subject.

A photovoltaic cell (PV cell) is a specialized semiconductor diode that converts visible light into direct current (DC). Some PV cells can also convert infrared (IR) or ultraviolet (UV) radiation into DC electricity. Photovoltaic cells are an integral part of solar-electric energy systems, which are becoming increasingly important as alternative sources of utility power.

The first PV cells were made of silicon combined, or doped, with other elements to affect the behavior of electrons or holes (electron absences within atoms). Other materials, such as copper indium diselenide (CIS), cadmium telluride (CdTe), and gallium arsenide (GaAs), have been developed for use in PV cells. There are two basic types of semiconductor material, called positive (or P type) and negative (or N type). In a PV cell, flat pieces of these materials are placed together, and the physical boundary between them is called the P-N junction. The device is constructed in such a way that the junction can be exposed to visible light, IR, or UV. When such radiation strikes the P-N junction, a voltage difference is produced between the P type and N type materials. Electrodes connected to the semiconductor layers allow current to be drawn from the device.

Large sets of PV cells can be connected together to form solar modules, arrays, or panels. The use of PV cells and batteries for the generation of usable electrical energy is known as photovoltaics. One of the major advantages of photovoltaics is the fact that it is non-polluting, requiring only real estate (and a reasonably sunny climate) in order to function. Another advantage is the fact that solar energy is unlimited. Once a photovoltaic system has been installed, it can provide energy at essentially no cost for years, and with minimal maintenance.

TYPES OF PHOTOVOLTAIC CELLS

Photovoltaic cells or PV cells can be manufactured in many different ways and from a variety of different materials. Despite this difference, they all perform the same task of harvesting solar energy and converting it to useful electricity. The most common material for solar panel construction is silicon which has semiconducting properties. Several of these solar cells are required to construct a solar panel and many panels make up a photovoltaic array.

There are three types of PV cell technologies that dominate the world market: monocrystalline silicon, polycrystalline silicon, and thin film. Higher efficiency PV technologies, including gallium arsenide and multi-junction cells, are less common due to their high cost, but are ideal for use in concentrated photovoltaic systems and space applications. There is also an assortment of emerging PV cell technologies which include Perovskite cells, organic solar cells, dye-sensitized solar cells and quantum dots.

A solar panel, consisting of many monocrystalline cells.

Monocrystalline Silicon Cell

The first commercially available solar cells were made from monocrystalline silicon, which is an extremely pure form of silicon. To produce these, a seed crystal is pulled out of a mass of molten silicon creating a cylindrical ingot with a single, continuous, crystal lattice structure. This crystal is then mechanically sawn into thin wafers, polished and doped to create the required p-n junction. After an anti-reflective coating and the front and rear metal contacts are added, the cell is finally wired and packaged alongside many other cells into a full solar panel. Monocrystalline silicon cells are highly efficient, but their manufacturing process is slow and labour intensive, making them more expensive than their polycrystalline or thin film counterparts.

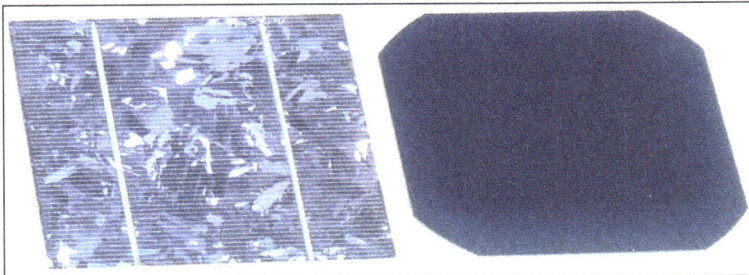

An image comparing a polycrystalline silicon cell (left) and a monocrystalline silicon cell (right).

Polycrystalline Silicon Cell

Instead of a single uniform crystal structure, polycrystalline (or multicrystalline) cells contain many small grains of crystals. They can be made by simply casting a cube-shaped ingot from molten silicon, then sawn and packaged similar to monocrystalline cells. Another method known as edge-defined film-fed growth (EFG) involves drawing a thin ribbon of polycrystalline silicon from a mass of molten silicon. A cheaper but less efficient alternative, polycrystalline silicon PV cells dominate the world market, representing about 70% of global PV production in 2015.

Thin Film Cells

Although crystalline PV cells dominate the market, cells can also be made from thin films—making them much more flexible and durable. One type of thin film PV cell is amorphous silicon (a-Si) which is produced by depositing thin layers of silicon on to a glass substrate. The result is a very thin and flexible cell which uses less than 1% of the silicon needed for a crystalline cell. Due to this reduction in raw material and a less energy intensive manufacturing process, amorphous silicon cells are much cheaper to produce. Their efficiency, however, is greatly reduced because the silicon atoms are much less ordered than in their crystalline forms leaving 'dangling bonds' that combine with other elements making them electrically inactive. These cells also suffer from a 20% drop in efficiency within the first few months of operation before stabilizing, and are therefore sold with power ratings based on their degraded output.

Other types of thin film cells include copper indium gallium diselenide (CIGS) and cadmium telluride (CdTe). These cell technologies offer higher efficiencies than amorphous silicon, but contain rare and toxic elements including cadmium which requires extra precautions during manufacture and eventual recycling.

A thin film solar panel composed of non-crystalline silicon deposited on a flexible material.

High Efficiency Cells

Other cell technologies have been developed which operate at much higher efficiencies than those mentioned above, but their higher material and manufacturing costs currently prohibit wide spread commercial use.

Gallium Arsenide

Silicon is not the only material suitable for crystalline PV cells. Gallium arsenide (GaAs) is an alternative semiconductor which is highly suitable for PV applications. Gallium arsenide has a similar crystal structure to that of monocrystalline silicon, but with alternating gallium and arsenic atoms.

Due to its higher light absorption coefficient and wider band gap, GaAs cells are much more efficient than those made of silicon. Additionally, GaAs cells can operate at much higher temperatures without considerable performance degradation, making them suitable for concentrated photovoltaics. GaAs cells are produced by depositing layers of gallium and arsenic onto a base of single crystal GaAs, which defines the orientation of the new crystal growth. This process makes GaAs cells much more expensive than silicon cells, making them useful only when high efficiency is needed, such as space applications.

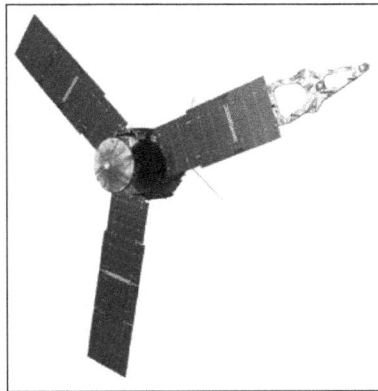

NASA's Juno Spacecraft with gallium arsenide multi-junction solar cells.

Multi-junction

The majority of PV cells contain only one p-n junction of semiconductor material which converts energy from one discreet portion of the solar spectrum into useful electricity. Multi-junction cells have 2 or more junctions layered on top of each other, allowing energy to be collected from multiple portions of the spectrum. Light that is not absorbed by the first layer will travel through and interact with subsequent layers. Multi-junction cells are produced in the same way as gallium arsenide cells—slowly depositing layers of material onto a single crystal base, making them very expensive to produce, and only commercially viable in concentrated PV systems and space applications.

Emerging Cell Technologies

Electricity can be produced through the interaction of light on many other materials as well. Perovskite solar cells, named after their specific crystal structure, can be produced from organic compounds of lead and elements such as chlorine, bromine or iodine. They are relatively cheap to produce and can boast efficiencies close to those of commercially available silicon cells but they are currently limited by a short lifespan. Organic solar cells consist of layers of polymers and can be produced cheaply at high volumes. These cells can be produced as a semi-transparent film, but suffer from relatively low efficiencies. Dye-sensitized solar cells can be produced using semiconducting titanium dioxide and a layer of 'sensitizer' dye only one molecule thick. These cells boast modest efficiencies but cannot withstand bright sunlight without degrading. Quantum dots utilize nanotechnology to manipulate semiconducting materials at extremely small scales. 'Nanoparticles' consisting of a mere 10,000 atoms can be tuned to different parts of the solar spectrum according to their size, and combined to absorb a wide range of energy. Although theoretical efficiencies are extremely high, laboratory test efficiencies are still very low.

PHOTOVOLTAIC CELL CHARACTERISTICS

Solar Cell I-V Characteristic Curves show the current and voltage (I-V) characteristics of a particular photovoltaic (PV) cell, module or array giving a detailed description of its solar energy conversion ability and efficiency. Knowing the electrical I-V characteristics (more importantly P_{max}) of a solar cell, or panel is critical in determining the device's output performance and solar efficiency.

Photovoltaic solar cells convert the suns radiant light directly into electricity. With increasing demand for a clean energy source and the sun's potential as a free energy source, has made solar energy conversion as part of a mixture of renewable energy sources increasingly important. As a result, the demand for efficient solar cells, which convert sunlight directly into electricity, is growing faster than ever before.

Photovoltaic (PV) cells are made made almost entirely from silicon that has been processed into an extremely pure crystalline form that absorbs the photons from sunlight and then releases them as electrons, causing an electric current to flow when the photoconductive cell is connected to an external load. There are a variety of different measurements we can make to determine the solar cell's performance, such as its power output and its conversion efficiency.

The main electrical characteristics of a PV cell or module are summarized in the relationship between the current and voltage produced on a typical solar cell I-V characteristics curve. The intensity of the solar radiation (insolation) that hits the cell controls the current (I), while the increases in the temperature of the solar cell reduces its voltage (V).

Solar cells produce direct current (DC) electricity and current times voltage equals power, so we can create solar cell I-V curves representing the current versus the voltage for a photovoltaic device.

Solar Cell I-V Characteristics Curves are basically a graphical representation of the operation of a solar cell or module summarising the relationship between the current and voltage at the existing conditions of irradiance and temperature. I-V curves provide the information required to configure a solar system so that it can operate as close to its optimal peak power point (MPP) as possible.

Solar Cell I-V Characteristic Curve

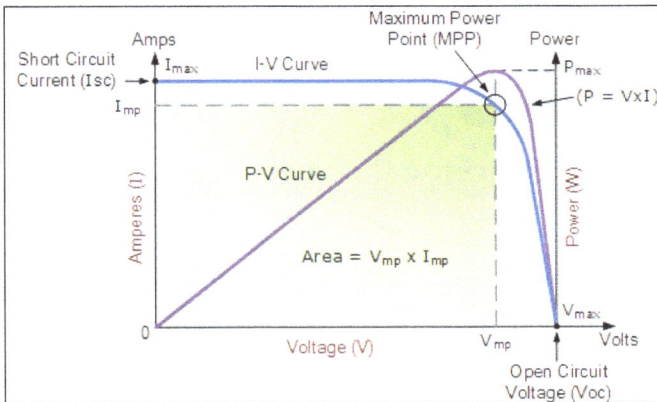

The above graph shows the current-voltage (I-V) characteristics of a typical silicon PV cell operating under normal conditions. The power delivered by a solar cell is the product of current and voltage (I x V). If the multiplication is done, point for point, for all voltages from short-circuit to open-circuit conditions, the power curve above is obtained for a given radiation level.

With the solar cell open-circuited, that is not connected to any load, the current will be at its minimum (zero) and the voltage across the cell is at its maximum, known as the solar cells open circuit voltage, or Voc. At the other extreme, when the solar cell is short circuited, that is the positive and negative leads connected together, the voltage across the cell is at its minimum (zero) but the current flowing out of the cell reaches its maximum, known as the solar cells short circuit current, or Isc.

Then the span of the solar cell I-V characteristics curve ranges from the short circuit current (Isc) at zero output volts, to zero current at the full open circuit voltage (Voc).

In other words, the maximum voltage available from a cell is at open circuit, and the maximum current at closed circuit. Of course, neither of these two conditions generates any electrical power, but there must be a point somewhere in between were the solar cell generates maximum power.

However, there is one particular combination of current and voltage for which the power reaches its maximum value, at Imp and Vmp. In other words, the point at which the cell generates maximum electrical power and this is shown at the top right area of the green rectangle. This is the "maximum power point" or MPP. Therefore the ideal operation of a photovoltaic cell (or panel) is defined to be at the maximum power point.

The maximum power point (MPP) of a solar cell is positioned near the bend in the I-V characteristics curve. The corresponding values of Vmp and Imp can be estimated from the open circuit voltage and the short circuit current: Vmp $\cong$ (0.8–0.90)Voc and Imp $\cong$ (0.85–0.95)Isc. Since solar cell output voltage and current both depend on temperature, the actual output power will vary with changes in ambient temperature.

Thus far we have looked at Solar Cell I-V Characteristic Curve for a single solar cell or panel. But many photovoltaic arrays are made up of smaller PV panels connected together. Then the I-V curve of a PV array is just a scaled up version of the single solar cell I-V characteristic curve as shown.

Solar Panel I-V Characteristic Curves

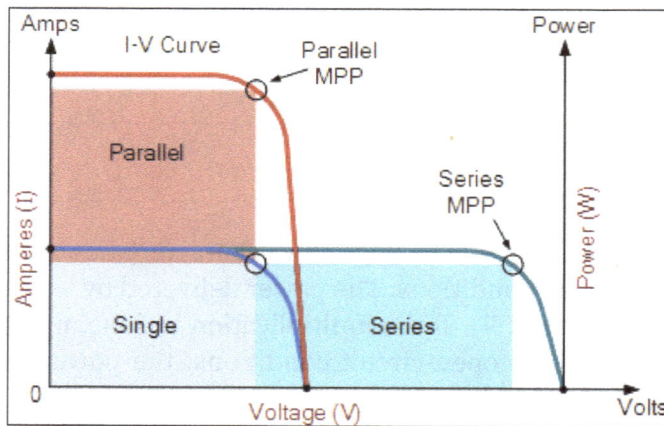

Photovoltaic panels can be wired or connected together in either series or parallel combinations, or both to increase the voltage or current capacity of the solar array. If the array panels are connected together in a series combination, then the voltage increases and if connected together in parallel then the current increases. The electrical power in Watts, generated by these different photovoltaic combinations will still be the product of the voltage times the current, (P = V x I). However the solar panels are connected together, the upper right hand corner will always be the maximum power point (MPP) of the array.

Electrical Characteristics of a Photovoltaic Array

The electrical characteristics of a photovoltaic array are summarised in the relationship between the output current and voltage. The amount and intensity of solar insolation (solar irradiance) controls the amount of output current (I), and the operating temperature of the solar cells affects the output voltage (V) of the PV array. Solar cell I-V characteristic curves that summarise the relationship between the current and voltage are generally provided by the panels manufacturer and are given as:

Solar Array Parameters

- V_{oc} = open-circuit voltage: This is the maximum voltage that the array provides when the terminals are not connected to any load (an open circuit condition). This value is much higher than Vmp which relates to the operation of the PV array which is fixed by the load. This value depends upon the number of PV panels connected together in series.

- I_{sc} = short-circuit current: The maximum current provided by the PV array when the output connectors are shorted together (a short circuit condition). This value is much higher than Imp which relates to the normal operating circuit current.

- MPP = maximum power point: This relates to the point where the power supplied by the array that is connected to the load (batteries, inverters) is at its maximum value, where MPP = Imp x Vmp. The maximum power point of a photovoltaic array is measured in Watts (W) or peak Watts (Wp).

- FF = fill factor: The fill factor is the relationship between the maximum power that the array can actually provide under normal operating conditions and the product of the open-circuit voltage times the short-circuit current, (Voc x Isc) This fill factor value gives an idea of the quality of the array and the closer the fill factor is to 1 (unity), the more power the array can provide. Typical values are between 0.7 and 0.8.

- %eff = percent efficiency: The efficiency of a photovoltaic array is the ratio between the maximum electrical power that the array can produce compared to the amount of solar irradiance hitting the array. The efficiency of a typical solar array is normally low at around 10-12%, depending on the type of cells (monocrystalline, polycrystalline, amorphous or thin film) being used.

Solar Cell I-V Characteristic Curves are graphs of output voltage versus current for different levels of insolation and temperature and can tell you a lot about a PV cell or panel's ability to convert sunlight into electricity. The most important values for calculating a particular panels power rating are the voltage and current at maximum power.

Some solar panels are rated at slightly higher or lower voltages than others of the same wattage value, and this affects the amount of current available and therefore the panels MPP. Other parameters also important are the open circuit voltage and short circuit current ratings from a safety point of view, especially the voltage rating. An array of six panels in series, while having a nominal 72 volt (6 x 12) rating, could potentially produce an open-circuit voltage of over 120 volts DC, which is more than enough to be dangerous.

Photovoltaic I-V characteristics curves provide the information needed for us to configure a solar power array so that it can operate as close as possible to its maximum peak power point. The peak power point is measured as the PV module produces its maximum amount of power when exposed to solar radiation equivalent to 1000 watts per square metre, 1000 W/m² or 1kW/m².

PHOTOVOLTAIC EFFECT

Photovoltaic effect is a process in which two dissimilar materials in close contact produce an electrical voltage when struck by light or other radiant energy. Light striking crystals such as silicon or germanium, in which electrons are usually not free to move from atom to atom within the crystal, provides the energy needed to free some electrons from their bound condition. Free electrons cross the junction between two dissimilar crystals more easily in one direction than in the other, giving one side of the junction a negative charge and, therefore, a negative voltage with respect to the other side, just as one electrode of a battery has a negative voltage with respect to the other. The photovoltaic effect can continue to provide voltage and current as long as light continues to fall on the two materials. This current can be used to measure the brightness of the incident light or as a source of power in an electrical circuit, as in a solar power system.

The photovoltaic effect in a solar cell can be illustrated with an analogy to a child at a slide. Initially, both the electron and the child are in their respective "ground states." Next, the electron is lifted up to its excited state by consuming energy received from the incoming light, just as the child is lifted up to an "excited state" at the top of the slide by consuming chemical energy stored in his body. In both cases there is now energy available in the excited state that can be expended. In the absence of junction-forming materials, there is no incentive for excited, free electrons to move along a specific direction; they eventually fall back to the ground state. On the other hand, whenever two different materials are placed in contact, an electric field is generated along the contact. This is the so-called built-in field, and it exerts a force on free electrons, effectively "tilting" the electron states and forcing the excited free electrons into an external electrical load where their excess energy can be dissipated. The external load can be a simple resistor, or it can be any of a myriad of electrical or electronic devices ranging from motors to radios. Correspondingly, the child moves to the slide because of his desire for excitement.

It is on the slide that the child dissipates his excess energy. Finally, when the excess energy is expended, both the electron and the child are back in the ground state, where they can begin the whole process over again. The motion of the electron, like that of the child, is in one direction, as can be seen from the figure. In short, the photovoltaic effect produces a direct current (DC)—one that flows constantly in only a single direction.

CONSTRUCTION OF PHOTOVOLTAIC CELL

It is a device which converts the light energy into electrical energy. When light is allowed to fall on this cell, the cell generates a voltage across its terminals. This voltage increase whit increase in the light intensity. The cell is so designed that a large area is exposed to light which enhances the voltage generation across the two terminals of the cell. Construction and working is explained below.

Construction and Working of Solar Cell

It essentially consists of a silicon PN junction diode with a glass window on top surface layer of P material is made extremely thin so, that incident light photon's may easily reach the PN junction. When these photons collide with valence electrons'. They comport them sufficient energy as to leave their parent atoms. In this way free electrons and holes are generated on both sides of the junction. Due to these holes and electrons current are produces. This current is directly proportional to the illumination's (mw/cm^2) and also depends on the size of the surface area being illuminated.

The open circuit voltage is a function of illumination. The symbol is shown below.

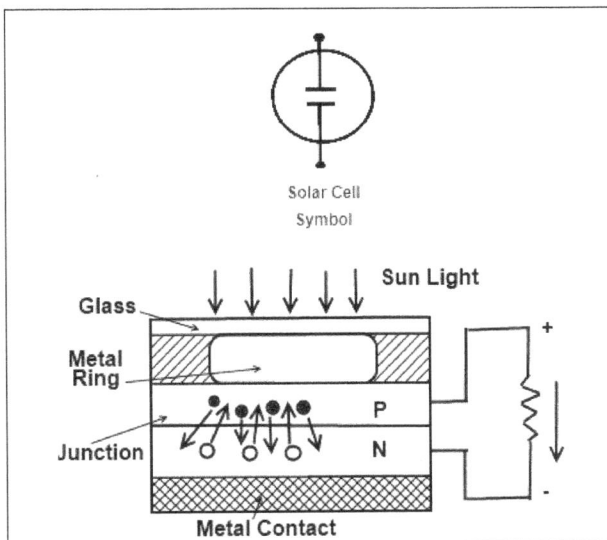

Solar Cell Construction.

As shown in the given diagram the Solar cell is like an ordinary diode. It consist of silicon, germanium PN junction with a glass windows on the top surface layer of P-Type, the P-Type material is made very thin and wide so that the incident light photon may easily reach to PN junction.

The P nickel plated ring around the P layer acts as the positive output terminal's (anode) and the metal contact at the bottom acts as a Cathode.

Silicon and germanium are the most widely used semiconductors materials for solar cells although Gallium arsenide, Indium arsenide and Cadmium arsenide are also being used nowadays.

WORKING OF PHOTOVOLTAIC CELL

Conversion of light energy in electrical energy is based on a phenomenon called photovoltaic effect. When semiconductor materials are exposed to light, the some of the photons of light ray are absorbed by the semiconductor crystal which causes a significant number of free electrons in the crystal. This is the basic reason for producing electricity due to photovoltaic effect. Photovoltaic cell is the basic unit of the system where the photovoltaic effect is utilised to produce electricity from light energy. Silicon is the most widely used semiconductor material for constructing the photovoltaic cell. The silicon atomhas four valence electrons. In a solid crystal, each silicon atom shares each of its four valence electrons with another nearest silicon atom hence creating covalent bonds between them. In this way, silicon crystal gets a tetrahedral lattice structure. While light ray strikes on any materials some portion of the light is reflected, some portion is transmitted through the materials and rest is absorbed by the materials.

The same thing happens when light falls on a silicon crystal. If the intensity of incident light is high enough, sufficient numbers of photons are absorbed by the crystal and these photons, in turn, excite some of the electrons of covalent bonds. These excited electrons then get sufficient energy to migrate from valence band to conduction band. As the energy level of these electrons is in the conduction band, they leave from the covalent bond leaving a hole in the bond behind each removed electron. These are called free electrons move randomly inside the crystal structure of the silicon. These free electrons and holes have a vital role in creating electricity in photovoltaic cell. These electrons and holes are hence called light-generated electrons and holesrespectively. These light generated electrons and holes cannot produce electricity in the silicon crystal alone. There should be some additional mechanism to do that.

When a pentavalent impurity such as phosphorus is added to silicon, the four valence electrons of each pentavalent phosphorous atom are shared through covalent bonds with four neighbour silicon atoms, and fifth valence electron does not get any chance to create a covalent bond.

This fifth electron then relatively loosely bounded with its parent atom. Even at room temperature, the thermal energy available in the crystal is large enough to disassociate these relatively loose fifth electrons from their parent phosphorus atom. While this fifth relatively loose electron is disassociated from parent phosphorus atom, the phosphorous atom immobile positive ions. The said disassociated electron becomes free but does not have any incomplete covalent bond or hole in the crystal to be re-associated. These free electrons come from pentavalent impurity are always ready to conduct current in the semiconductor. Although there are numbers of free electrons, still the substance is electrically neutral as the number of positive phosphorous ions locked inside the crystal structure is exactly equal to the number of the free electrons come out from them. The process of inserting impurities in the semiconductor is known as doping, and the impurities are doped are known as dopants. The pentavalent dopants which donate their fifth free electron to the semiconductor crystal are known as donors. The semiconductors doped by donor impurities are known as n-type or negative type semiconductor as there are plenty of free electrons which are negatively charged by nature.

When instead pentavalent phosphorous atoms, trivalent impurity atoms like boron are added to a semiconductor crystal opposite type of semiconductor will be created. In this case, some silicon atoms in the crystal lattice will be replaced by boron atoms, in other words, the boron atoms will occupy the positions of replaced silicon atoms in the lattice structure. Three valance electrons of boron atom will pair with valance electron of three neighbour silicon atoms to create three complete covalent bonds. For this configuration, there will be a silicon atom for each boron atom, fourth valence electron of which will not find any neighbour valance electrons to complete its fourth covalent bond. Hence this fourth valence electron of these silicon atoms remains unpaired and behaves as incomplete bond. So there will be lack of one electron in the incomplete bond, and hence an incomplete bond always attracts electron to fulfil this lack. As such, there is a vacancy for the electron to sit.

This vacancy is conceptually called positive hole. In a trivalent impurity doped semiconductor, a significant number of covalent bonds are continually broken to complete other incomplete covalent bonds. When one bond is broken one hole is created in it. When one bond is completed, the hole in it disappears. In this way, one hole appears to disappear another neighbour hole. As such holes are having relative motion inside the semiconductor crystal. In the view of that, it can be said holes also can move freely as free electrons inside semiconductor crystal. As each of the holes can accept an electron, the trivalent impurities are known as acceptor dopants and the semiconductors doped with acceptor dopants are known as p-type or positive type semiconductor.

In n-type semiconductor mainly the free electrons carry negative charge and in p-type semiconductor mainly the holes in turn carry positive charge therefore free electrons in n-type semiconductor and free holes in p-type semiconductor are called majority carrier in n-type semiconductor and p-type semiconductor respectively.

There is always a potential barrier between n-type and p-type material. This potential barrier is essential for working of a photovoltaic or solar cell. While n-type semiconductor and p-type semiconductor contact each other, the free electrons near to the contact surface of n-type semiconductor get plenty of adjacent holes of p-type material. Hence free electrons in n-type semiconductor near to its contact surface jump to the adjacent holes of p-type material to recombine. Not only free electrons, but valence electrons of n-type material near the contact surface also come out from the covalent bond and recombine with more nearby holes in the p-type semiconductor. As the co-valent bonds are broken, there will be a number of holes created in the n-type material near the contact surface. Hence, near contact zone, the holes in the p-type materials disappear due to recombination on the other hand holes appear in the n-type material near same contact zone. This is as such equivalent to the migration of holes from p-type to the n-type semiconductor. So as soon as one n-type semiconductor and one p-type semiconductor come into contact the electrons from n-type will transfer to p-type and holes from p-type will transfer to n-type. The process is very fast but does not continue forever. After some instant, there will be a layer of negative charge (excess electrons) in the p-type semiconductor adjacent to the contact along the contact surface. Similarly, there will be a layer of positive charge (positive ions) in the n-type semiconductor adjacent to contact along the contact surface. The thickness of these negative and positive charge layer increases up to a certain extent, but after that, no more electrons will migrate from n-type semiconductor to p-type semiconductor. This is because, while any electron of n-type semiconductor tries to migrate over p-type semiconductor it faces a sufficiently thick layer of positive ions in n-type semiconductor itself where it will drop without crossing it. Similarly, holes will no more migrate to n-type semiconductor from p-type. The holes when trying to cross the negative layer in p-type semiconductor these will recombine with electrons and no more movement toward n-type region.

In other words, negative charge layer in the p-type side and positive charge layer in n-type side together form a barrier which opposes migration of charge carriers from its one side to other. Similarly, holes in the p-type region are held back from entering the n-type region. Due to positive and negative charged layer, there will be an electric field across the region and this region is called depletion layer.

Now let us come to the silicon crystal. When light ray strikes on the crystal, some portion of the light is absorbed by the crystal, and consequently, some of the valence electrons are excited and come out from the covalent bond resulting free electron-hole pairs.

If light strikes on n-type semiconductor the electrons from such light-generated electron-hole pairs are unable to migrate to the p-region since they are not able to cross the potential barrier due to the repulsion of an electric field across depletion layer. At the same time, the light-generated holes cross the depletion region due to the attraction of electric field of depletion layer where they recombine with electrons, and then the lack of electrons here is compensated by valence electrons of p-region, and this makes as many numbers of holes in the p-region. As such light generated holes are shifted to the p-region where they are trapped because once they come to the p-region cannot be able to come back to n-type region due to the repulsion of potential barrier.

As the negative charge (light generated electrons) is trapped in one side and positive charge (light generated holes) is trapped in opposite side of a cell, there will be a potential difference between these two sides of the cell. This potential difference is typically 0.5 V. This is how a photovoltaic cells or solar cells produce potential difference.

APPLICATION OF PHOTOVOLTAIC CELL

A photovoltaic system, or solar PV system is a power system designed to supply usable solar power by means of photovoltaics. It consists of an arrangement of several components, including solar panels to absorb and directly convert sunlight into electricity, a solar inverter to change the electric current from DC to AC, as well as mounting, cabling and other electrical accessories. PV systems range from small, roof-top mounted or building-integrated systems with capacities from a few to several tens of kilowatts, to large utility-scale power stations of hundreds of megawatts. Nowadays, most PV systems are grid-connected, while stand-alone systems only account for a small portion of the market.

Rooftop and Building Integrated Systems

Rooftop PV on half-timbered house.

Photovoltaic arrays are often associated with buildings: either integrated into them, mounted on them or mounted nearby on the ground. Rooftop PV systems are most often retrofitted into existing buildings, usually mounted on top of the existing roof structure or on the existing walls. Alternatively, an array can be located separately from the building but connected by cable to supply power for the building. Building-integrated photovoltaics (BIPV) are increasingly incorporated into the roof or walls of new domestic and industrial buildings as a principal or ancillary source of electrical power. Roof tiles with integrated PV cells are sometimes used as well. Provided there is an open gap in which air can circulate, rooftop mounted solar panels can provide a passive cooling effect on buildings during the day and also keep accumulated heat in at night. Typically, residential rooftop systems have small capacities of around 5–10 kW, while commercial rooftop systems often amount to several hundreds of kilowatts. Although rooftop systems are much smaller than ground-mounted utility-scale power plants, they account for most of the worldwide installed capacity.

Concentrator Photovoltaics

Concentrator photovoltaics (CPV) is a photovoltaic technology that contrary to conventional flat-plate PV systems uses lenses and curved mirrors to focus sunlight onto small, but highly efficient, multi-junction (MJ) solar cells. In addition, CPV systems often use solar trackers and sometimes a cooling system to further increase their efficiency. Ongoing research and development is rapidly improving their competitiveness in the utility-scale segment and in areas of high solar insolation.

Photovoltaic Thermal Hybrid Solar Collector

Photovoltaic thermal hybrid solar collector (PVT) are systems that convert solar radiation into thermal and electrical energy. These systems combine a solar PV cell, which converts sunlight into electricity, with a solar thermal collector, which captures the remaining energy and removes waste heat from the PV module. The capture of both electricity and heat allow these devices to have higher exergy and thus be more overall energy efficient than solar PV or solar thermal alone.

Power Stations

Satellite image of the Topaz Solar Farm.

Many utility-scale solar farms have been constructed all over the world. As of 2015, the 579-megawatt (MW_{AC}) Solar Star is the world's largest photovoltaic power station, followed by the Desert Sunlight Solar Farm and the Topaz Solar Farm, both with a capacity of 550 MW_{AC}, constructed by US-company First Solar, using CdTe modules, a thin-film PV technology. All three power stations are located in the Californian desert. Many solar farms around the world are integrated with agriculture and some use innovative solar tracking systems that follow the sun's daily path across the sky to generate more electricity than conventional fixed-mounted systems. There are no fuel costs or emissions during operation of the power stations.

Rural Electrification

Developing countries where many villages are often more than five kilometers away from grid power are increasingly using photovoltaics. In remote locations in India a rural lighting program has been providing solar powered LED lighting to replace kerosene lamps. The solar powered lamps were sold at about the cost of a few months' supply of kerosene.Cuba is working to provide solar power for areas that are off grid. More complex applications of off-grid solar energy use include 3D printers. RepRap 3D printers have been solar powered with photovoltaic technology,which enables distributed manufacturing for sustainable development. These are areas where the social costs and benefits offer an excellent case for going solar, though the lack of profitability has relegated such endeavors to humanitarian efforts. However, in 1995 solar rural electrification projects had been found to be difficult to sustain due to unfavorable economics, lack of technical support, and a legacy of ulterior motives of north-to-south technology transfer.

Standalone Systems

Until a decade or so ago, PV was used frequently to power calculators and novelty devices. Improvements in integrated circuits and low power liquid crystal displays make it possible to power such devices for several years between battery changes, making PV use less common. In contrast, solar powered remote fixed devices have seen increasing use recently in locations where significant connection cost makes grid power prohibitively expensive. Such applications include solar lamps, water pumps,parking meters,emergency telephones, trash compactors,temporary traffic signs, charging stations,and remote guard posts and signals.

Floating Solar

Where land may be limited, PV can be deployed as floating solar. In May 2008, the Far Niente Winery in Oakville, CA pioneered the world's first "floatovoltaic" system by installing 994 photovoltaic solar panels onto 130 pontoons and floating them on the winery's irrigation pond. The floating system generates about 477 kW of peak output and when combined with an array of cells located adjacent to the pond is able to fully offset the winery's electricity consumption. The primary benefit of a floating system is that it avoids the need to sacrifice valuable land area that could be used for another purpose. In the case of the Far Niente Winery, the floating system saved three-quarters of an acre that would have been required for a land-based system. That land area can instead be used for agriculture.Another benefit of a floating solar system is that the panels are kept at a lower temperature than they would be on land, leading to a higher efficiency of solar energy conversion. The floating panels also reduce the amount of water lost through evaporation and inhibit the growth of algae.

In Transport

Solar Impulse 2, a solar aircraft.

PV has traditionally been used for electric power in space. PV is rarely used to provide motive power in transport applications, but is being used increasingly to provide auxiliary power in boats and cars. Some automobiles are fitted with solar-powered air conditioning to limit interior temperatures on hot days. A self-contained solar vehicle would have limited power and utility, but a solar-charged electric vehicle allows use of solar power for transportation. Solar-powered cars, boatsand airplaneshave been demonstrated, with the most practical and likely of these being solar cars. The Swiss solar aircraft, Solar Impulse 2, achieved the longest non-stop solo flight in history and completed the first solar-powered aerial circumnavigation of the globe in 2016.

Telecommunication and Signaling

Solar PV power is ideally suited for telecommunication applications such as local telephone exchange, radio and TV broadcasting, microwave and other forms of electronic communication links. This is because, in most telecommunication application, storage batteries are already in use and the electrical system is basically DC. In hilly and mountainous terrain, radio and TV signals may not reach as they get blocked or reflected back due to undulating terrain. At these locations, low power transmitters (LPT) are installed to receive and retransmit the signal for local population.

Spacecraft Applications

Part of Juno's solar array.

Solar panels on spacecraft are usually the sole source of power to run the sensors, active heating and cooling, and communications. A battery stores this energy for use when the solar panels are in shadow. In some, the power is also used for spacecraft propulsion—electric propulsion. Spacecraft were one of the earliest applications of photovoltaics, starting with the silicon solar cells used on the Vanguard 1 satellite, launched by the US in 1958. Since then, solar power has been used on missions ranging from the MESSENGER probe to Mercury, to as far out in the solar system as the Juno probe to Jupiter. The largest solar power system flown in space is the electrical system of the International Space Station. To increase the power generated per kilogram, typical spacecraft solar panels use high-cost, high-efficiency, and close-packed rectangular multi-junction solar cells made of gallium arsenide (GaAs) and other semiconductor materials.

Specialty Power Systems

Photovoltaics may also be incorporated as energy conversion devices for objects at elevated temperatures and with preferable radiative emissivities such as heterogeneous combustors.

ADVANTAGES AND DISADVANTAGES OF PHOTOVOLTAIC CELL

Advantages of Solar PV

- PV panels provide clean – green energy. During electricity generation with PV panels there is no harmful greenhouse gas emissions thus solar PV is environmentally friendly.

- Solar energy is energy supplied by nature – it is thus free and abundant.

- Solar energy can be made available almost anywhere there is sunlight.

- Solar energy is especially appropriate for smart energy networks with distributed power generation – DPG is indeed the next generation power network structure.

- Solar panels cost is currently on a fast reducing track and is expected to continue reducing for the next years – consequently solar PV panels has indeed a highly promising future both for economical viability and environmental sustainability.

- Photovoltaic panels, through photoelectric phenomenon, produce electricity in a direct electricity generation way.

- Operating and maintenance costs for PV panels are considered to be low, almost negligible, compared to costs of other renewable energy systems.

- PV panels have no mechanically moving parts, except in cases of sun-tracking mechanical bases; consequently they have far less breakages or require less maintenance than other renewable energy systems (e.g. wind turbines).

- PV panels are totally silent, producing no noise at all; consequently, they are a perfect solution for urban areas and for residential applications.

- Because solar energy coincides with energy needs for cooling, PV panels can provide an effective solution to energy demand peaks – especially in hot summer months where energy demand is high.

- Though solar energy panels' prices have seen a drastic reduction in the past years, and are still falling, nonetheless, solar photovoltaic panels are one of major renewable energy systems that are promoted through government subsidy funding (FITs, tax credits etc.); thus financial incentive for PV panels make solar energy panels an attractive investment alternative.

- Residential solar panels are easy to install on rooftops or on the ground without any interference to residential lifestyle.

Disadvantages of Solar PV

- As in all renewable energy sources, solar energy has intermittency issues; not shining at night but also during daytime there may be cloudy or rainy weather.

- Consequently, intermittency and unpredictability of solar energy makes solar energy panels less reliable a solution.

- Solar energy panels require additional equipment (inverters) to convert direct electricity (DC) to alternating electricity (AC) in order to be used on the power network.

- For a continuous supply of electric power, especially for on-grid connections, Photovoltaic panels require not only inverters but also storage batteries; thus increasing the investment cost for PV panels considerably.

- In case of land-mounted PV panel installations, they require relatively large areas for deployment; usually the land space is committed for this purpose for a period of 15-20 years – or even longer.

- Solar panels efficiency levels are relatively low (between 14%-25%) compared to the efficiency levels of other renewable energy systems.

- Though PV panels have no considerable maintenance or operating costs, they are fragile and can be damaged relatively easily; additional insurance costs are therefore of ultimate importance to safeguard a PV investment.

References

- Photovoltaic-cell-PV-Cell: whatis.techtarget.com, Retrieved 19 April, 2019

- Types-of-photovoltaic-cells, encyclopedia: energyeducation.ca, Retrieved 08 August, 2019

- Solar-cell-i-v-characteristic, energy-articles: alternative-energy-tutorials.com, Retrieved 10 January, 2019

- Photovoltaic-effect, science: britannica.com, Retrieved 19 July, 2019

- Working-principle-of-photovoltaic-cell-or-solar-cell: electrical4u.com, Retrieved 17 May, 2019

- Pathak, M. J. M.; Sanders, P. G.; Pearce, J. M. (2014). "Optimizing limited solar roof access by exergy analysis of solar thermal, photovoltaic, and hybrid photovoltaic thermal systems". Applied Energy. 120: 115–124. CiteSeerX 10.1.1.1028.406. doi:10.1016/j.apenergy.2014.01.041

- Advantages-and-disadvantages-of-solar-photovoltaic-quick-pros-and-cons-of-solar-pv: renewableenergyworld.com, Retrieved 25 June, 2019

5

Phototransistors

An electronic component which is either used for current switching or current amplification is defined as phototransistor. It relies on light and operates in reverse bias connection. This chapter delves into the study of characteristics, applications and advantages and disadvantages of phototransistors to provide in-depth knowledge of the subject.

The phototransistor is a three-layer semiconductor device which has a light-sensitive base region. The base senses the light and converts it into the current which flows between the collector and the emitter region.

The construction of phototransistor is similar to the ordinary transistor, except the base terminal. In phototransistor, the base terminal is not provided, and instead of the base current, the light energy is taken as the input.

Symbol of Phototransistor

The symbol of the phototransistor is similar to that of the ordinary transistor. The only difference is that of the two arrows which show the light incident on the base of the phototransistor.

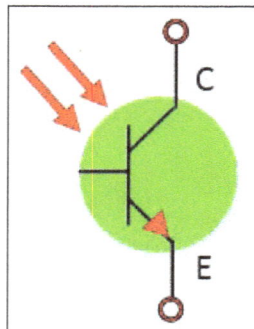

Principle of Phototransistor

Consider the conventional transistor is having open terminal base circuited. The collector base leakage current acts as a base current I_{CBO}.

$$I_C = \beta I_B + (1+B)\, I_{CBO}$$

As the base current $I_B = 0$, it acts as an open circuited. And the collector current becomes.

$$I_C = (1+B) I_{CBO}$$

The above equations shown that the collector current is directly proportional to the current base leakage current, i.e., the I_C increases with the increases of the collector base region.

Phototransistor Operation

The phototransistor is made up of semiconductor material. When the light was striking on the material, the free electrons/holes of the semiconductor material causes the current which flows in the base region. The base of the phototransistor would only be used for biasing the transistor. In case of NPN transistor, the collector is made positive concerning emitter, and in PNP, the collector is kept negative.

The light enters into the base region of phototransistor generates the electron-hole pairs. The generation of electron-hole pairs mainly occurs into the reverse biasing. The movement of electrons under the influence of electric field causes the current in the base region. The base current injected the electrons in the emitter region. The major drawback of the phototransistor is that they have low-frequency response.

Phototransistor Construction

The construction of the phototransistor is quite similar to the ordinary transistor. Earlier, the germanium and silicon are used for fabricating the phototransistor. The small hole is made on the surface of the collector-base junction for placing the lens. The lens focuses the light on the surface.

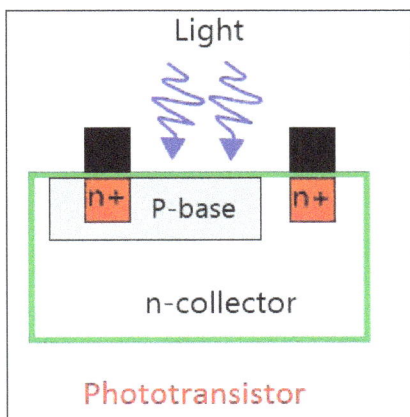

Nowadays the transistor is made of a highly light effective material (like gallium and arsenides). The emitter-base junction is kept at forward biased, and the collector-base junction is at the reverse biased.

When no light falls on the surface of the transistor, the small reverse saturation current induces on the transistor. The reverse saturation current induces because of the few minority charge carriers. The light energy falls on the collector-base junction and generates the more majority charge carrier which adds the current to the reverse saturation current. The graph below shows the magnitude of current increases along with the intensity of light.

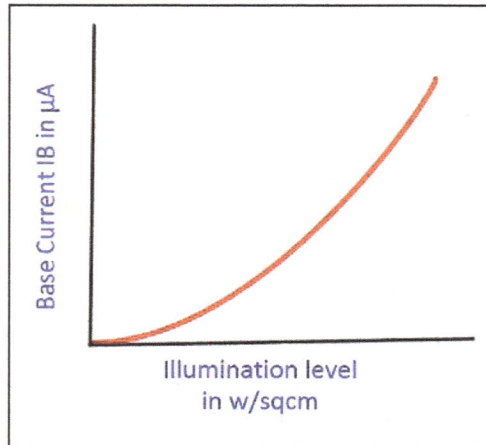

The phototransistor is widely used in electronics devices likes smoke detectors, infrared receiver, CD players, lasers etc. for sensing light.

CHARACTERISTICS OF PHOTOTRANSISTORS

An equivalent circuit for a phototransistor consists of a photodiode feeding its output photocurrent into the base of a small signal transistor. Based on this model it is not surprising that phototransistors display some of the characteristics of both types of devices.

Spectral Response

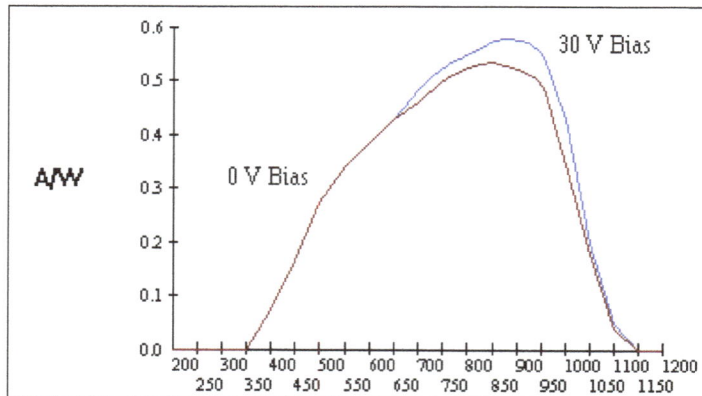

Phototransistors will respond to fluorescent or incandescent light sources but display better optical coupling efficiencies when matched with IR LEDs. Standard IR LEDs are GaAs (940 nm) and GaAlAs (880 nm).

The output of a phototransistor is dependent upon the wavelength of incident light. These devices respond to light over a broad range of wavelengths from the near UV, through the visible and into the near IR part of the spectrum. Unless optical filters are used, the peak spectral response is in the near IR at approximately 840 nm. The peak response is at a somewhat shorter wavelength than that of a typical photodiode. This is because the diffused junctions of a phototransistor are formed in epitaxial rather than crystal grown silicon wafers.

Sensitivity

For a given light source illumination level, the output of a phototransistor is defined by the area of the exposed collector-base junction and the dc current gain of the transistor. The collector-base junction of the phototransistor functions as a photodiode generating a photocurrent which is fed into the base of the transistor section. Thus, like the case for a photodiode, doubling the size of the base region doubles the amount of generated base photocurrent. This photocurrent (I_P) then gets amplified by the dc current gain of the transistor. For the case where no external base drive current is applied:

$$I_C = h_{FE} (I_P)$$

where:

I_C = collector current

h_{FE} = DC current gain

I_P = photocurrent

As is the case with signal transistors, h_{FE} is not a constant but varies with base drive, bias voltage and temperature. At low light levels the gain starts out small but increases with increasing light (or base drive) until a peak is reached. As the light level is further increased the gain of the phototransistor starts to decrease.

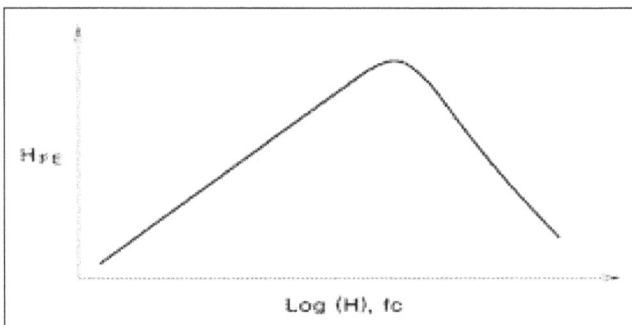

Transistor Gain vs Light Intensity.

H_{FE} will also increase with increasing values for V_{CE}. The current -voltage characteristics of a typical transistor will demonstrate this effect.

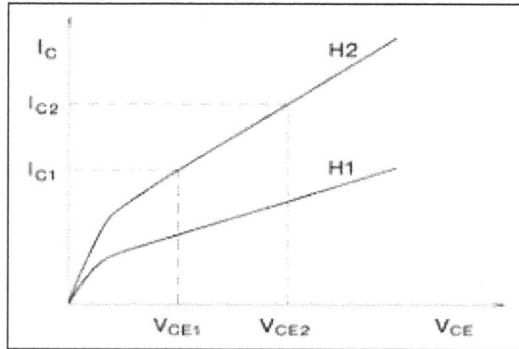

Current vs Voltage Curves.

For a constant base drive the curve shows a positive slope with increasing voltage. It is clear that the current gain at collector-emitter voltage V_{CE2} is greater than the current gain at V_{CE1}. The current gain will also increase with increasing temperature.

Linearity

Unlike a photodiode whose output is linear with respect to incident light over 7 to 9 decades of light intensity, the collector current (I_C) of a phototransistor is linear for only 3 to 4 decades of illumination. The prime reason for this limitation is that the dc gain (h_{FE}) of the phototransistor is a function of collector current (I_C) which in turn is determined by the base drive. The base drive may be in the form of a base drive current of incident light.

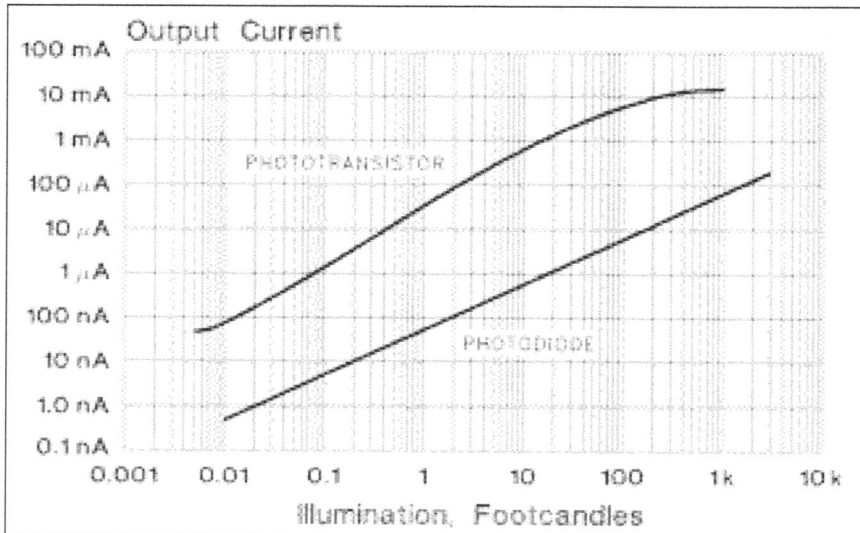

Photodetector Relative Linearity.

While photodiodes are the detector of choice when linear output versus light intensity is extremely important, as in light intensity measuring equipment, the phototransistor comes into its own when the application requires a photodetector to act like a switch.

When light is present, a phototransistor or photodarlington can be considered "on", a condition during which they are capable of sinking a fair amount of current. When the light is removed these photodetectors enter an "off" state and function electrically as open switches.

Collector-emitter Saturation Voltage - $V_{CE(SAT)}$

Saturation is the condition in which both the emitter-base and the collector-base junctions of a phototransistor become forward biased. From a practical standpoint the collector-emitter saturation voltage, $V_{CE(SAT)}$, is the parameter which indicates how closely the photodetector approximates a closed switch. This is because $V_{CE(SAT)}$ is the voltage dropped across the detector when it is in its "on" state.

$V_{CE(SAT)}$ is usually given as the maximum collector-emitter voltage allowed at a given light intensity and for a specified value of collector current. EG&G Optoelectronics tests their detectors for $V_{CE(SAT)}$ at a light level of 400 fc and with 1 mA of collector current flowing through the device. Stock phototransistors are selected according to a set of specifications where $V_{CE(SAT)}$ can range from 0.25 V (max) to 0.55 V (max) depending on the device.

Dark Current - (I_D)

When the phototransistor is placed in the dark and a voltage is applied from collector to emitter, a certain amount of current will flow. This current is called the dark current (I_D). This current consists of the leakage current of the collector-base junction multiplied by the dc current gain of the transistor. The presence of this current prevents the phototransistor from being considered completely "off", or being an ideal "open" switch.

The dark current is specified as the maximum collector current permitted to flow at a given collector-emitter test voltage. The dark current is a function of the value of the applied collector-emitter voltage and ambient temperature.

EG&G Optoelectronics stock phototransistors and photodarlingtons are tested at a V_{CE} applied voltage of either 5 V, 10 V or 20 V depending on the device. Phototransistors are tested to dark current limits which range from 10 nA to 100 nA.

Dark current is temperature dependent, increasing with increasing temperature. It is usually specified at 25°C.

Breakdown Voltages - (V_{BR})

Phototransistors must be properly biased in order to operate. However, when voltages are applied to the phototransistor, care must be taken not to exceed the collector-emitter breakdown voltage (V_{BRCEO}) or the emitter-collector breakdown voltage (V_{BRECO}). Exceeding the breakdown voltages cause permanent damage to the phototransistor.

Typical values for V_{BRCEO} range from 20 V to 50 V. Typical values for V_{BRECO} range from 4 V to 6 V. The breakdown voltages are 100% screened parameters.

Speed of Response

The speed of response of a phototransistor is dominated almost totally by the capacitance of the collector-base junction and the value of the load resistance. These dominate due to the Miller Effect which multiplies the value of the RC time constant by the current gain of the phototransistor. This leads to the general rule that for devices with the same active area, the higher the gain of the photodetector, the slower will be its speed of response.

A phototransistor takes a certain amount of time to respond to sudden changes in light intensity. This response time is usually expressed by the rise time (t_R) and fall time (t_F) of the detector where:

t_R - The time required for the output to rise from 10% to 90% of its on-state value.
t_F - The time required for the output to fall from 90% to 10% of its on-state value.

As long as the light source driving the phototransistor is not intense enough to cause optical saturation, characterized by the storage of excessive amounts of charge carriers in the base region, risetime equals falltime. If optical saturation occurs, t_F can become much larger than t_R.

EG&G Optoelectronics tests the t_R and t_F of its phototransistors and photodarlingtons at an I_C = 1.0 mA and with a 100 ohm load resistor in series with the detector. Phototransistors display t_R and t_F times in a range of 1 μsec to 10 μsec. The t_R and t_F of photodarlingtons typically lie between 30 μsec and 250 μsec.

Selecting a Photodetector

Each application is a unique combination of circuit requirements, light intensity levels, wavelengths, operating environment and cost considerations. EG&G Optoelectronics offers a broad range of catalog phototransistors and photodarlingtons to help you with these design tradeoffs.

The charts presented below are intended to give some general guidelines and tradeoffs for selecting the proper detector for your application.

Size of Detector Chip		
Small Size	Parameter	Large Size
Lower	Sensitivity	Higher
Faster	Speed of Response	Slower
Lower	Dark Current	Higher
Lower	Cost	Higher

Gain (H_{FE})		
Low Gain	Parameter	High Gain
Lower	Sensitivity	Higher
Faster	Speed of Response	Slower
Lower	Dark Current	Higher
Smaller	Temp. Coef.	Larger
Lower	Cost	Higher

APPLICATIONS OF PHOTOTRANSISTORS

The applications of phototransistors are as follows:

- In light controlling and detection: As phototransistors are a very sensitive light detector. Thus these are widely used in light detection and controlling applications.

- In an indication of level and relays: The device finds its uses in indicating the level of some systems because of their light sensing ability.

- In counting systems: Phototransistors can be effectively utilized in counting systems. As it has tremendous ability to combinely operate as photodiode and transistors. Thus, failure of supply will not cause much adverse effects on the system.

- In punch card readers: Phototransistors widely finds its applications in punch card reading.

ADVANTAGES AND DISADVANTAGES OF PHOTOTRANSISTORS

Advantages of Phototransistor

- There are numerous applications of phototransistor such as light measurement, light sensitive switch, opto-coupler etc.

- The dark current of a phototransistor is much higher compare to photodiode.

- It has ability to sink current of about 20 mA to 50 mA. Hence it can be connected to component having low impedance. For example, phototransistor can drive piezoelectric audio transducer or LED indicator directly. This means it can directly interface to small loads.

- Phototransistor is more responsive to light than photoresistor.

- Phototransistors are cheaper.

- Phototransistors can produce a voltage which can be used in conjunction with microcontroller housing ADC.

- Output current of phototransistor is easily obtainable.

Disadvantages of Phototransistor

- It is insensitive to incident light from other directions than particular narrow window unlike photoresistor which is sensitive to incident light from anywhere in front of it.

- Silicon phototransistors do not handle higher voltages above 1000V.

- They are more susceptible to electricity surges/spikes and EM energy from radiations.

- It has nonlinear characteristic and it is temperature sensitive.

- Large dispersion between individual units.

6

Photoresistors

Photoresistor is a device that exhibits photoconductivity by decreasing the resistance with increasing intensity of light on its sensitive surface. It can be categorized into intrinsic photoresistor and extrinsic photoresistor. This chapter sheds light on these types of photoresistors and their applications for providing a thorough understanding of the subject.

Photoresistor is the combination of words "photon" (meaning light particles) and "resistor". True to its name, a photo-resistor is a device or we can say a resistor dependent on the light intensity. For this reason, they are also known as light dependent a.k.a. LDRs.

From our basic knowledge about the relationship between resistivity (ability to resist the flow of electrons) and conductivity (ability to allow the flow of electrons), we know that both are polar opposites of each other. Thus when we say that the resistance decreases when intensity of light increases, it simply implies that the conductance increases with increase in intensity of light falling on the photo-resistor or the LDR, owing to a property called photo-conductivity of the material.

Hence these Photoresistors are also known as photoconductive cells or just photocell.

The idea of Photoresistor developed when photoconductivity in Selenium was discovered by Willoughby Smith in 1873. Many variants of the photoconductive devices were then made.

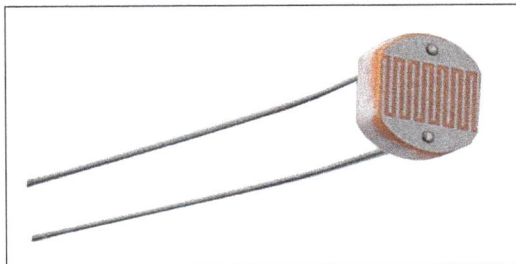

Photoresistor

Photoresistor Symbol

In order to represent a Photoresistor in a circuit diagram, the symbol chosen was that would indicate it to be a light dependent device along with the fact that it is a resistor.

While mostly the symbol used is shown in figure (two arrows pointing to a resistor), some prefer to encase the resistor in a circle like that shown in figure.

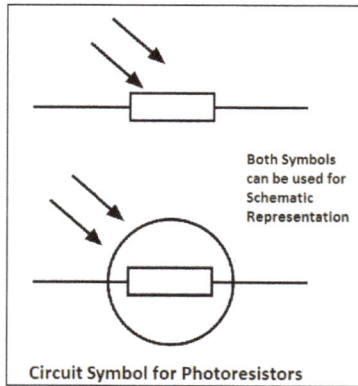

Photoresistors Circuit Symbol.

Working Principle of a Photoresistor

In order to understand the working principle of a Photoresistor, let's brush up a little about the valence electrons and the free electrons.

As we know valence electrons are those found in the outermost shell of an atom. Hence, these are loosely attached to the nucleus of the atom. This means that only some small amount of energy is needed to pull it out from the outer orbit.

Free electrons on the other hand are those which are not attached to the nucleus and hence free to move when an external energy like an electric field is applied. Thus when some energy makes the valence electron pull out from the outer orbit, it acts as a free electron; ready to move whenever an electric field is applied. The light energy is used to make valence electron a free electron.

This very basic principle is used in the Photoresistor. The light that falls on a photoconductive material is absorbed by it which in turn makes lots of free electrons from the valence electrons.

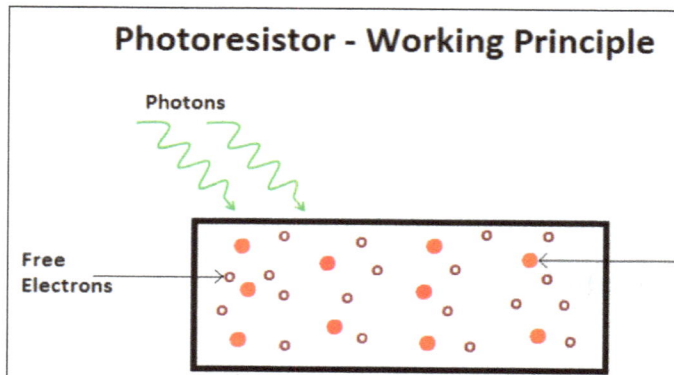

Photoresistor Working Principle

As the light energy falling on the photoconductive material increases, number of valence electrons that gain energy and leave the bonding with the nucleus increases. This leads to a large number of valence electrons jump to the conduction band, ready to move with an application of any external force like an electric field.

Thus, as the light intensity increases, the number of free electrons increases. This means the photoconductivity increases that imply a decrease in photo resistivity of the material.

Now that we have covered the working mechanism, we got an idea that a photoconductive material is used for the construction of a Photoresistor. According to the type of photoconductive material the Photoresistors are of two types.

Basic Photoresistor Circuit

The figure below shows a basic circuit diagram of a Photoresistor ciruit. It has a battery, a Photoresistor and a led. This setup helps understand the behaviour of Photoresistor when subjected to an electric field.

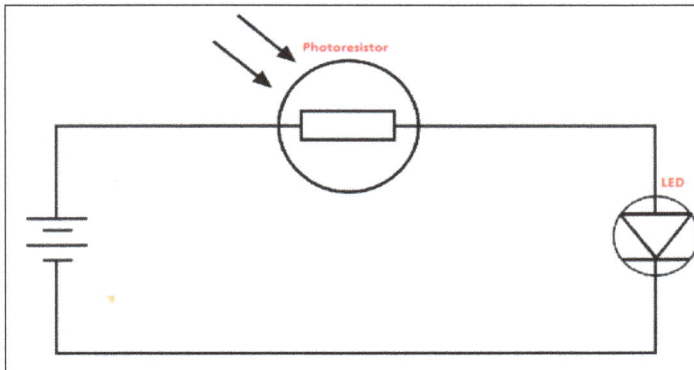

Basic Photoresistor Circuit.

Photoresistor – Uses and Applications

Automatic Street Lights: One of the prominent uses of Photoresistor that we experience in daily life is in the circuits of automatic street lights, as already hinted in the introductory paragraph. Here they are so used in a circuit that the street lights turn on as it starts getting dark and turns off in the morning.

Some of the Photoresistors are used in some of the consumer items like light meters in camera, light sensors like in robotic projects, clock radios etc.

They are also used to control the reduction in gain of dynamic compressors.

They are also considered as a good infra-red detector and hence find application in infrared astronomy.

Applications

Most common application in the circuits of automatic street lights, and other consumer items like light meter, light sensor etc.

TYPES OF PHOTORESISTORS

Intrinsic Photoresistors

Intrinsic photoresistors are made from the pure semiconductor materials such as silicon or germanium. The outermost shell of any atom is capable to hold up to eight valence electrons. However, in silicon or germanium, each atom consists of only four valence electrons. These four valence electrons of each atom form four covalent bonds with the neighboring four atoms to completely fill the outermost shell. As a result, no electron is left free.

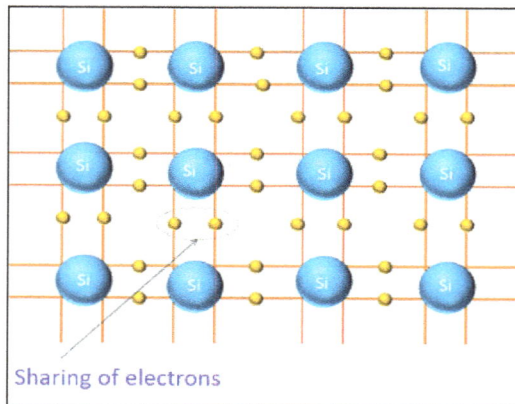

Sharing of electrons

When we apply light energy to the intrinsic photo resistor, only a small number of valence electrons gain enough energy and becomes free from the parent atom. Hence, a small number of charge carriers are generated. As a result, only a small electric current flows through the intrinsic photo resistor.

We already have known that increase in electric current means decrease in resistance. In intrinsic photoresistors, the resistance decreases slightly with the increase in light energy. Hence, intrinsic photoresistors are less sensitive to the light. Therefore, they are not reliable for the practical applications.

Extrinsic Photoresistors

Extrinsic photoresistors are made from the extrinsic semiconductor materials. Let us consider an example of extrinsic photoresistor, which is made from the combination of silicon and impurity (phosphorus) atoms.

Each silicon atom consists of four valence electrons and each phosphorus atom consists of five valence electrons. The four valence electrons of the phosphorus atom form four covalent bonds with the neighboring four silicon atoms. However, the fifth valence electron of the phosphorus atom cannot able to form the covalent bond with the silicon atom because the silicon atom has only four valence electrons. Hence, the fifth valence electron of each phosphorus atom becomes free from the atom. Thus, each phosphorus atom generates a free electron.

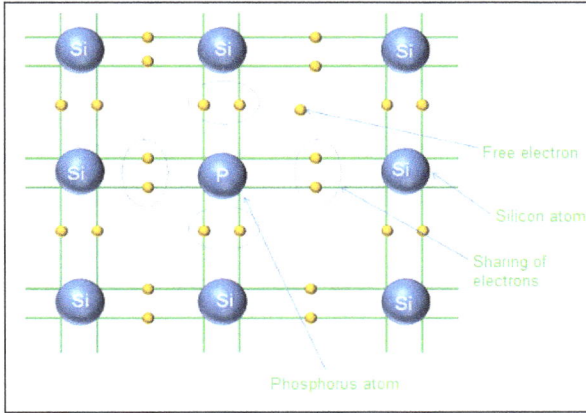

The free electron, which is generated will collides with the valence electrons of other atoms and makes them free. Likewise, a single free electron generates multiple free electrons. Therefore, adding a small number of impurity (phosphorus) atoms generates millions of free electrons.

In extrinsic photoresistors, we already have large number of charge carriers. Hence, providing a small amount of light energy generates even more number of charge carriers. Thus, the electric current increases rapidly.

Increase in electric current means decrease in resistance. Therefore, the resistance of the extrinsic photoresistor decreases rapidly with the small increase in applied light energy. Extrinsic photoresistors are reliable for the practical applications.

CHARACTERISTICS OF A PHOTORESISTOR

Wavelength Dependency

The sensitivity of a photo resistor varies with the light wavelength. If the wavelength is outside a certain range, it will not affect the resistance of the device at all. It can be said that the LDR is not sensitive in that light wavelength range. Different materials have different unique spectral response curves of wavelength versus sensitivity. Extrinsic light dependent resistors are generally designed for longer wavelengths of light, with a tendency towards the infrared (IR). When working in the IR range, care must be taken to avoid heat buildup, which could affect measurements by changing the resistance of the device due to thermal effects. The figure shown here represents the spectral response of photoconductive detectors made of different materials, with the operating temperature expressed in K and written in the parentheses.

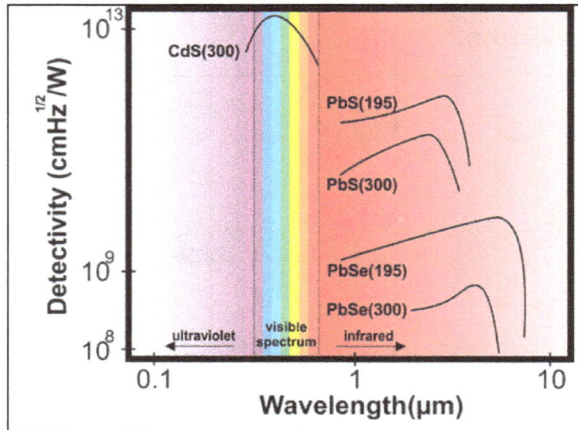

Sensitivity

Light dependent resistors have a lower sensitivity than photo diodes and photo transistors. Photo diodes and photo transistors are true semiconductor devices which use light to control the flow of electrons and holes across PN-junctions, while light dependent resistors are passive components, lacking a PN-junction. If the light intensity is kept constant, the resistance may still vary significantly due to temperature changes, so they are sensitive to temperature changes as well. This property makes LDRs unsuitable for precise light intensity measurements.

Latency

Another interesting property of photo resistors is that there is time latency between changes in illumination and changes in resistance. This phenomenon is called the resistance recovery rate. It takes usually about 10 ms for the resistance to drop completely when light is applied after total darkness, while it can take up to 1 second for the resistance to rise back to the starting value after the complete removal of light. For this reason the LDR cannot be used where rapid fluctuations of light are to be recorded or used to actuate control equipment, but this same property is exploited in some other devices, such as audio compressors, where the function of the light dependent resistor is to smooth the response.

Construction and Properties of Photoresistors

Since the discovery of selenium photoconductivity, many materials have been found with similar characteristics. In the 1930s and 1940s PbS, PbSe and PbTe were studied following the development of photoconductors made of silicon and germanium. Modern light dependent resistors are made of lead sulfide, lead selenide, indium antimonide, and most commonly cadmium sulfide and cadmium selenide. The popular cadmium sulfide types are often indicated as CdS photoresistors. To manufacture a cadmium sulfide LDR, highly purified cadmium sulfide powder and inert binding materials are mixed. This mixture is then pressed and sintered. Electrodes are vacuum evaporated onto the surface of one side to form interleaving combs and connection leads are connected. The disc is then mounted in a glass envelope or encapsulated in transparent plastic to prevent surface contamination. The spectral response curve of cadmium sulfide matches that of the human eye. The peak sensitivity wavelength is about 560-600 nm which is in the visible part of the spectrum. It should be noted that devices containing lead or cadmium are not RoHS compliant and are banned for use in countries that adhere to RoHS laws.

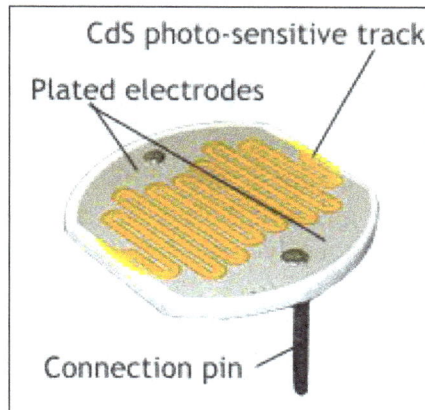

ADVANTAGES AND DISADVANTAGES OF PHOTORESISTOR

Advantages of Photoresistor

- Small in size.

- Low cost.

- It is easy to carry from one place to another place.

Disadvantages of Photoresistor

- The accuracy of photoresistor is very low.

7

Thermal Photodetectors

Thermal photodetectors or biometers are sensors of light with a p-n junction to convert photons into current. Terahertz photodetectors, bolometers, microbolometer, pyroelectric detectors, etc. fall under its domain. This chapter delves into the subject of thermal photodetectors for an extensive understanding of it.

TERAHERTZ PHOTODETECTORS

Terahertz (THz) range of electromagnetic spectrum still presents a challenge for both electronic and photonic technologies and is often described as the final unexplored area of spectrum. This radiation is frequently treated as the spectral region within frequency range $(v) \approx 0.1 - 10$ THz.

The past 20 years have seen a revolution in THz systems, as advanced materials research provided new and higher-power sources, and the potential of THz for advanced physics research and commercial applications was demonstrated. Numerous recent breakthroughs in the field have pushed THz research into the centre stage. As examples of milestone achievements can be included the development of THz time-domain spectroscopy (TDS), THz imaging, and high-power THz generation by means of nonlinear effects.

General Classification of Terahertz Detectors

The majority of detectors can be classified in two broad categories: photon detectors and thermal detectors.

Photon Detectors

In photon detectors the radiation is absorbed within the material by interaction with electrons either bound to lattice atoms or to impurity atoms or with free electrons. The radiation can be also absorbed by electrons localized in the quantum wells or in the mini-bands of superlattices. The observed electrical output signal results from the changed electronic energy distribution. The photon detectors show a selective wavelength dependence of response per unit incident radiation power.

Depending on the nature of the interaction, the class of photon detectors is further sub-divided into different types. The most important are: intrinsic detectors, extrinsic detectors, and photoemissive (Schottky barriers). Different types of detectors are described in detail in the monograph Infrared Detectors. Figure shows spectral detectivity characteristics of different types of detectors.

Photodetectors that utilize excitation of an electron from the valence to conduction band are called intrinsic detectors. Those which operate by exciting electrons into the conduction band or holes into the valence band from impurity states within the band (impurity-bound states in energy gap, quantum wells or quantum dots), are called extrinsic detectors. In comparison with intrinsic photoconductivity, the extrinsic photoconductivity is far less efficient because of limits in the amount of impurity that can be introduced into semiconductor without altering the nature of the impurity states. Intrinsic detectors are most common at the short wavelengths, below 20 μm.

A key difference between intrinsic and extrinsic detectors is that extrinsic detectors require much cooling to achieve high sensitivity at a given spectral response cutoff in comparison with intrinsic detectors. Low-temperature operation is associated with longer-wavelength sensitivity in order to suppress noise due to thermally induced transitions between close-lying energy levels. The long wavelength cutoff can be approximated as $T_{max} = 300\,K\,/\,\lambda_c[\mu m]$, where λ_c is the cut-off wavelength.

Thermal Detectors

The second class of detectors is composed of thermal detectors their operation principles are briefly described in Table. In a thermal detector the incident radiation is absorbed to change the material temperature, and the resultant change in some physical property is used to generate an electrical output. The detector is suspended on lags, which are connected to the heat sink. The signal does not depend upon the photonic nature of the incident radiation. Thus, thermal effects are generally wavelength independent; the signal depends upon the radiant power (or its rate of change) but not upon its spectral content. Attention is directed toward three approaches which have found the greatest utility in infrared technology, namely, bolometers, pyroelectric, and thermoelectric effects. In pyroelectric detectors a change in the internal electrical polarization is measured, whereas in the case of thermistor bolometers a change in the electrical resistance is measured.

Bolometers may be divided into several types. The most commonly used are the metal, the thermistor, and the semiconductor bolometers. A fourth type is the superconducting bolometer. This bolometer operates on a conductivity transition in which the resistance changes dramatically over the transition temperature range.

Comparison of the D* of various available detectors when operated at the indicated temperature. Chopping frequency is 1000 Hz for all detectors except the thermopile (10 Hz), thermocouple (10 Hz), thermistor bolometer (10 Hz), Golay cell (10 Hz) and pyroelectric detector (10 Hz). Each detector is assumed to view a surrounding hemisphere (2π field of view) at a temperature of 300 K. Theoretical curves for the background limited D* (dashed lines) for ideal photovoltaic and photoconductive detectors and thermal detectors are also shown. PC, photoconductive detector; PEM, photoelectronmagnetic detector; PV, photovoltaic detector; HEB, hot electron bolometer.

The key trade-off with respect to conventional thermal detectors is between sensitivity and response time. The detector sensitivity is often expressed by noise equivalent temperature (NEDT) represented by the temperature change, for incident radiation, that gives an output signal equal to the rms noise level. The thermal conductance is an extremely important parameter, since the noise equivalent temperature difference (NEDT) is proportional to $(G_{th})^{1/2}$, but the thermal response time of the detector, τ_{th}, is inversely proportional to G_{th}. Therefore, a change in thermal conductance due to improvements in material processing techniques improves sensitivity at the expense of time response.

Detectors for Room Temperature THz Imaging

Particular attention in development of THz imaging systems is devoted to the realization of sensors with a large potential for real-time imaging while maintaining a high dynamic range and room-temperature operation. CMOS process technology is especially attractive due to their low price tag for industrial, surveillance, scientific, and medical applications. However, CMOS THz imagers developed thus far have mainly operated single detectors based on lock-in technique to acquire raster-scanned imagers with frame rates on the order of minutes. With this mind, much of recent developments are directed towards three types of focal plane arrays (FPAs):

- Schottky barrier diodes (SBDs) compatible with the CMOS process,

- Field effect transistors (FETs) relying on plasmonic rectification phenomena,

- Adaptation of infrared bolometers to the THz frequency range.

An important issue for a FPA is pixel uniformity. It appears however, that the production of monolithically-integrated detector arrays encounters so many technological problems that the device-to-device performance variations and even the percentage of non-functional detectors per chip tend to be unacceptably high.

Mode of Operation: Thermopile

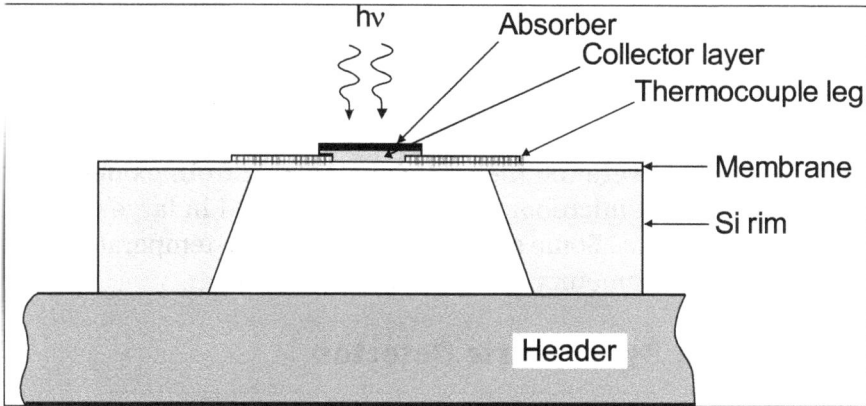

Operation and Properties

The thermocouple is usually a thin, blackened flake connected thermally to the junction of two dissimilar metals or semiconductors. Heat absorbed by the flake causes a temperature rise of the junction, and hence a thermoelectric electromotive force is developed which can be measured. Although thermopiles are not as sensitive as bolometers and pyroelectric detectors, they will replace these in many applications due to their reliable characteristics and good cost/performance ratio. Thermocouples are widely used in spectroscopy.

Mode of Operation: Bolometer Metal Semiconductor

Operation and Properties

The bolometer is a resistive element constructed from a material with a very small thermal capacity and large temperature coefficient so that the absorbed radiation produces a large change in resistance. The change in resistance is like the photoconductor;

however, the basic detection mechanisms are different. In the case of a bolometer, radiant power produces heat within the material, which in turn produces the resistance change. There is no direct photonelectron interaction.

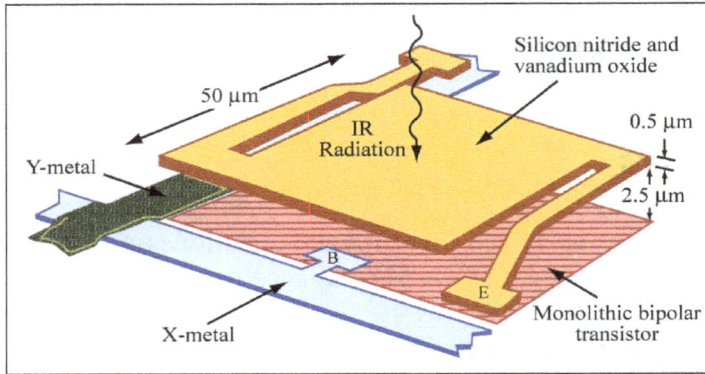

Mode of Operation: Superconductor Hot Electron

Operation and Properties

Initially, most bolometers were the thermistor type made from oxides of manganese, cobalt, or nickel. At present microbolometers are fabricated in large format arrays for thermal imaging applications. Some extremely sensitive low-temperature semiconductor and superconductor bolometers are used in THz region.

Mode of Operation: Pyroelectric Detector

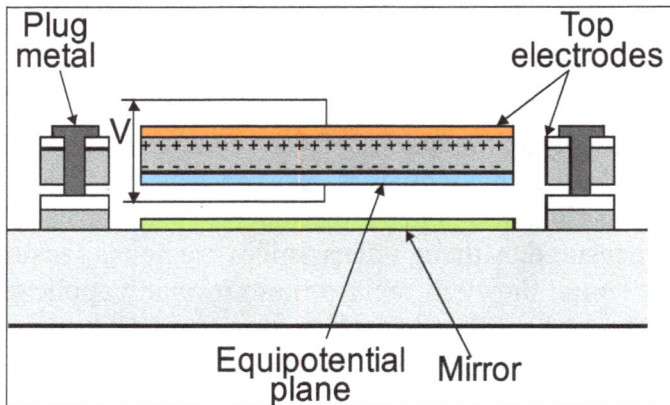

Operation and Properties

The pyroelectric detector can be considered as a small capacitor with two conducting electrodes mounted perpendicularly to the direction of spontaneous polarization. During incident of radiation, the change in polarization appears as a charge on the capacitor and a current is generated, the magnitude of which depends on the temperature rise and the pyroelectrical coefficient of the material. The signal, however, must be chopped or modulated. The detector sensitivity is limited either by amplifier noise or by loss-tangent noise. Response speed can be engineered making pyroelectric detectors useful for fast laser pulse detection, however with proportional decrease in sensitivity.

Mode of Operation: Golay Cell

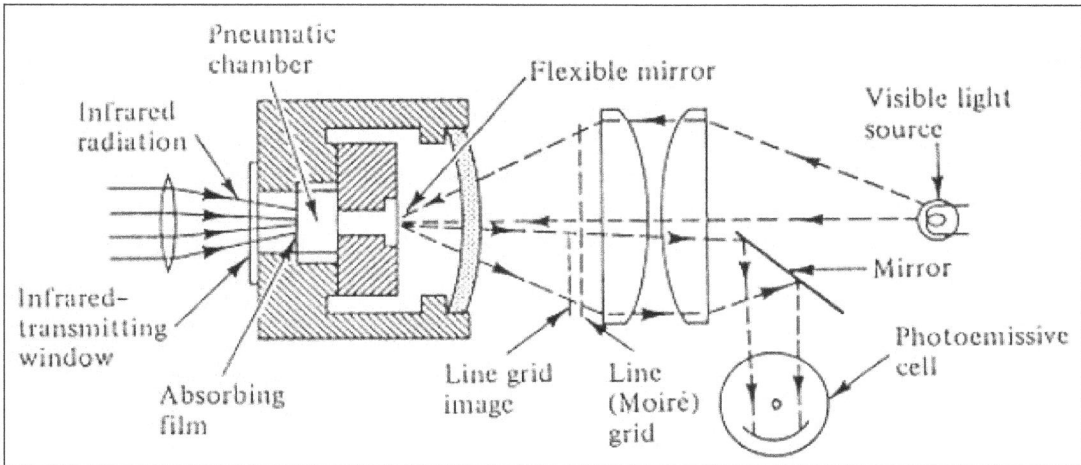

Operation and Properties

The Golay cell consists of an hermetically sealed container filled with gas (usually xenon for its low thermal conductivity) and arranged so that expansion of the gas under heating by a photon signal distorts a flexible membrane on which a mirror is mounted. The movement of the mirror is used to deflect a beam of light shining on a photocell and so producing a change in the photocell current as the output. In modern Golay cells the photocell is replaced by a solid state photodiode and light emitting diode is used for illumination.

The performance of the Golay cell is only limited by the temperature noise associated with the thermal exchange between the absorbing film and the detector gas; consequently, the detector can be extremely sensitive with $D^* \approx 3 \times 10^9$ cm Hz$^{1/2}$ W^{-1}, and responsivities of 10^5 to 10^6 V/W. The response time is quite long, typically 15 ms.

SPDs respond to the THz electric field and usually generate an output current or voltage through a quadratic term in their currentvoltage characteristics. In general, the noise equivalent power (NEP) of SBD and FET detectors is better than that of Golay cells and pyroelectric detectors around 300 GHz. Both the pyroelectric and the bolometer FPAs with detector response times in the millisecond time range are not suited for heterodyne operation. FET detectors are clearly capable in heterodyne detection with improving sensitivity. Diffraction aspects predicts FPAs for higher frequencies (0.5 THz and above) and in conjunction with large f/#optics.

Schottky Barrier Diodes

In spite of achievements of other kind of detectors for THz waveband, the Schottky barrier diodes (SBDs) are among the basic elements in THz technologies. They are used either in direct detection and as nonlinear elements in heterodyne receiver mixers operating in temperature range of 4–300 K. The cryogenically cooled SBDs were used in mixers preferably in 1980s and early 1990s and then they have been replaced widely by superconductor-insulator-superconductor (SIS) or hot electron bolometer (HEB) mixers, in which mixing processes are similar to that observed in SBDs, but, e.g. in SIS structures the rectification process is based on quantum-mechanical photon-assisted tunnelling of quasiparticles (electrons). The nonlinearity of SBD I–V characteristic (the current increases exponentially with the applied voltage) is the prerequisite for mixing to occur.

Historically first Schottky-barrier structures were pointed contacts of tapered metal wires (e.g. a tungsten needle) with a semiconductor surface (the so-called crystal detectors). Due to limitation of whisker technology, such as constraints on design and repeatability, starting in the 1980s, the efforts were made to produce planar Schottky diodes with air-bridge fingers. This design has been the most important steps toward a practical Schottky diode mixer for THz frequency applications, with several thousand diodes on a single chip and where parasitic losses such as the series resistance and the shunt capacitance are minimized. To achieve good performance at high frequencies the diode area should be small. Reducing junction area one reduces junction capacitances to increase operating frequency. But at the same time one increases the series resistance.

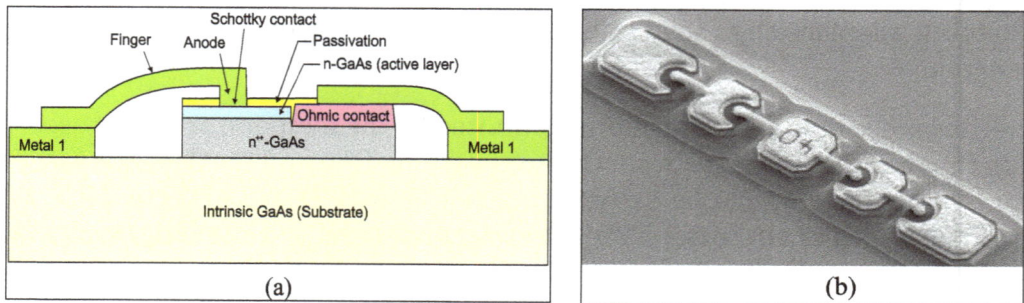

Gaas Schottky barrier diode: (a) schematic of a planar diode and (b) a four-diode chip array.

Using advanced technology, the diodes are integrated with many passive circuit elements (impedance matching, filters, and waveguide probes) onto the same substrate. By improving the mechanical arrangement and reducing loss, the planar technology is pushed well beyond 300 GHz up to several THz. For example, the figure below shows photographs of a bridged four-Schottky diodes' chip arrayed in a balanced configuration to increase power handling.

Recently, an alternative method of Schottky barrier formation has been elaborated by molecular beam epitaxy (MBE) in-situ deposition of a semimetal (ErAs) on semiconductor (InGaAs/InAlAs on InP substrates) to reduce the imperfections that give rise to excess low-frequency noise, particularly l/f noise. Excellent NEP performance for this III−V semiconductor SBD has been reported (1.4 pW/Hz$^{1/2}$ at 100 GHz). By using interband tunnelling, a heterojunction backward diode demonstrated 49.7 kV/W responsivity and 0.18 pW/Hz NEP at 94 GHz. More recently Han et al. have demonstrated fully functional CMOS imager operating near or in the sub-millimeter-wave frequency range. The 4 × 4 array increases the imaging speed by 4−8 times, due to fewer mechanical scan steps.

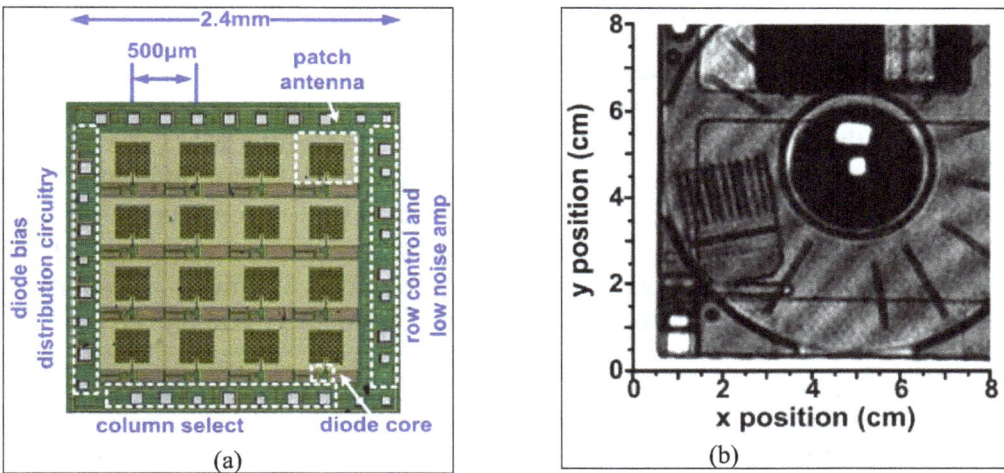

CMOS SBD 280-GHz imager: die photos of the array (a) and an image of music greeting card obtained. After Han, R., Zhang, Y., Kim, Y., et al., 2012. 280 GHz and 860 GHz image sensors using Schottky-barrier diodes in 0.13 μm digital CMOS. IEEE International Solid-State Circuits Conference, 254−256.

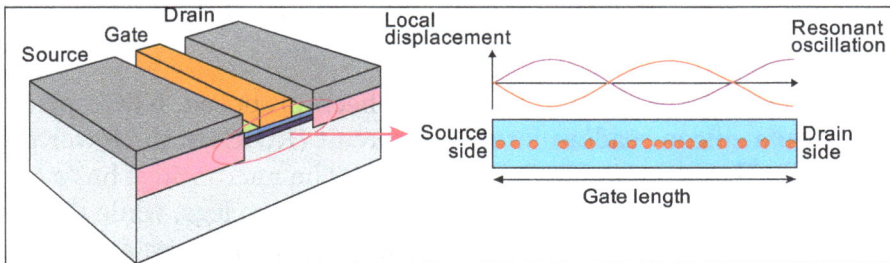

Plasma oscillations in a field effect transistor.

Field Effect Transistor and CMOS-based Detectors

The use field effect transistors (FETs) as detectors of THz radiation was first proposed by Dyakonov and Shur on the basis of formal analogy between the equations of the electron transport in a gated two-dimensional transistor channel and those of shallow water, or acoustic waves in music instruments. Figure schematically shows the resonant oscillation of plasma waves in gated region of FET.

The detection by FETs is due to nonlinear properties of the transistor, which lead to the rectification of the ac current induced by the coming radiation. As a result, a photoresponse appears in the form of a dc voltage between source and drain. This voltage is proportional to the radiation intensity (photovoltaic effect). A big improvement in sensitivity can be obtained by adding a proper antenna or a cavity coupling. The FETs can be used both for resonant (resonant case of high electron mobility, when plasma oscillation modes are excited in the channel) and non-resonant (broadband) THz detection in the case of low mobility, where plasma oscillations are overdamped.

The large-scale interest in using FETs as THz detectors started around 2004 after the first experimental demonstration of subTHz and THz detection in silicon-CMOS FETs.

Recently, the first CMOS FPA used to capture transmission-mode THz video streams in real-time without the need for raster scanning and source modulation has been fabricated. A camera with 32 × 32 pixel array fully integrated in a 65-nm CMOS process technology has been demonstrated. Each 80-mm array pixel consists of a differential on-chip ring antenna coupled to NMOS direct detector operated well-beyond its cut-off frequency. The camera chip has been packed together with a 41.7-dBi silicon lens in a 5×5×3 cm3 camera module. In continuous-wave illumination the camera achieves a responsivity of 100–200 kV/W and a total NEP of 10–20 nW/Hz$^{1/2}$ up to 500 fps at 856 GHz.

Microbolometers

An impressive promising technology is also coming from commercially available microbolometer arrays. Adaptation of infrared microbolometers to the THz frequency range after the successful demonstration of active THz imaging in 2006 entailed that in the period 2010–2011 three different companies/organizations announced cameras optimized for the 41-THz frequency range: NEC and Leti. The number of vendors is expected to increase soon.

Different designs of THz bolometer pixels have been proposed. NEC's pixel is divided into two parts: a silicon readout integrated circuit (ROIC) in the lower part, and a suspended microbridge structure in the upper part. The microbridge has a two-storied structure. The bottom is composed of a diaphragm and two legs, while the top (eaves) structure is formed on the diaphragm to increase the sensitive area and fill factor. The diaphragm and the eaves absorb THz radiation. The diaphragm is composed of VOx bolometer thin film, SiNx passivation layers and TiAlV electrodes, while the eaves structure is composed of SiNx layer and TiAlV thin film THz absorption layer.

A schematic of one Leti pixel of an amorphous silicon microbolometer array is shown in top centre of the figure. The 50-µm pitch is associated with quasi-double-bowtie antennas to a thermometer microbridge structure derived from the standard IR bolometer. The membrane is suspended over the substrate by arms and pillars. In order to enhance the antenna gain, an equivalent quarter-wavelength resonant cavity is realized under antennas with an 11-µm thick SiO2 layer deposited over the metallic reflector. To ensure electric contact between the bolometer pillars and CMOS metal upper contacts, the vias are etched through an 11-µm cavity and then metalized.

The NEP values for bolometer FPAs fabricated by three vendors. The FPAs optimized for 2–5 THz exhibit impressive NEP values below 100 pW/Hz$^{1/2}$. It can be seen that wavelength dependence of NEP is quite flat below 200 mm. Further improvement of performance is possible by increasing number of pixels, modification of antenna design while preserving pixel pitch, ROIC and technological stack.

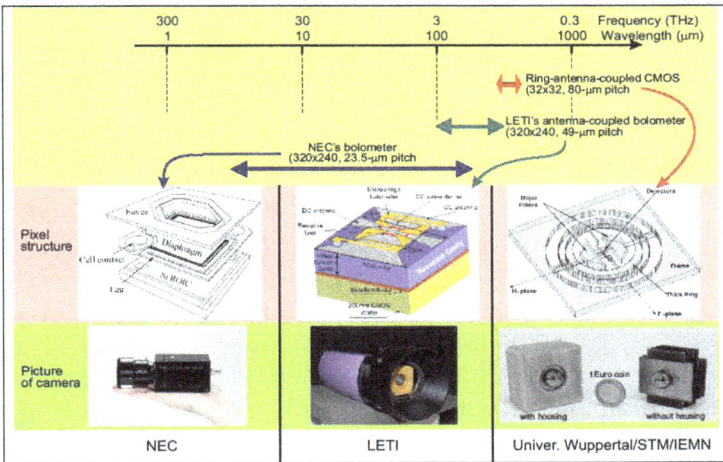

Development status of uncooled THz focal plane arrays.

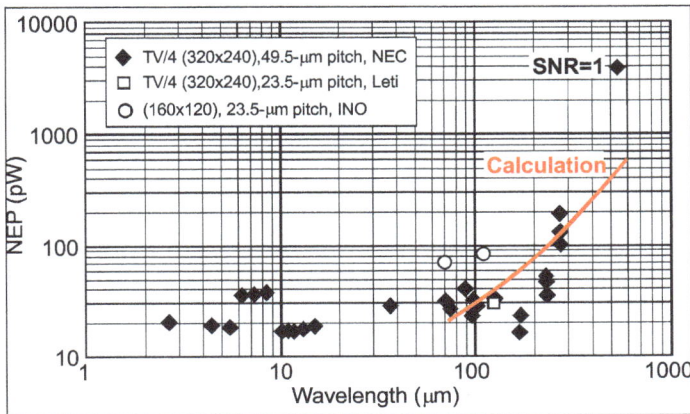

Spectral dependence of NEP for bolometer THz FPAs.

Extrinsic Semiconductor Detectors

Research and development of extrinsic IR photodetectors have been ongoing for more than 50 years. In the 1950s and 1960s, germanium could be made purer than silicon. Today, the problems with producing pure Si have been largely solved. Si has several advantages over Ge; for example, three orders of magnitude higher impurity solubilities are attainable, hence thinner detectors with better spatial resolution can be fabricated from silicon. Si has a lower dielectric constant than Ge, and the related device technology of Si has now been more thoroughly developed, including contacting methods, surface passivation, and mature MOS and CCD technologies. Moreover, Si detectors are characterized by superior hardness in nuclear radiation environments.

For wavelengths longer than 40 μm there are no appropriate shallow dopants for silicon; therefore, germanium devices are still of interest for very long wavelengths. Germanium photoconductors have been used in a variety of infrared astronomical experiments, both airborne and space-based at wavelength ranging from 3 to more than 200 μm. Very shallow donors, such as Sb, and acceptors, such as B, In, or Ga, provide cut-off wavelengths in the region of 100 μm. Figure shows the spectral response of the extrinsic germanium and silicon photoconductors.

Ge:Ga photoconductors are the best low background photon detectors for the wavelength range from 40 to 120 μm. Application of uniaxial stress along the axis of Ge:-Ga crystals reduces the Ga acceptor binding energy, extending the cutoff wavelength to ≈240μm. At the same time, the operating temperature must be reduced to less than 2 K.

The standard planar hybrid architecture, commonly used to construct near and mid-infrared focal-plane arrays, is not suitable for far IR detectors where readout glow, lack of efficient heat dissipation, and thermal mismatch between the detector and the readout could potentially limit their performance. Usually, the far-infrared arrays have a modular design with many modules stacked together to form a 2-dimensional array.

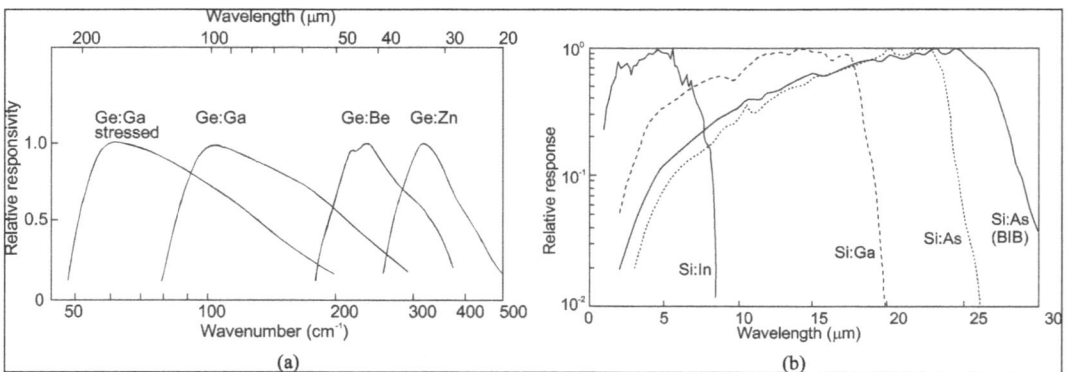

Relative spectral response of some germanium (a) and silicon (b) extrinsic photoconductors. For comparison also response of Si:As BIB is shown.

The Infrared Astronomical Satellite (IRAS), the Infrared Space Observatory (ISO), and for the far-infrared channels the SpitzerSpace Telescope (Spitzer) have all used bulk germanium photoconductors. In the Spitzer mission a 32 × 32-pixel Ge:Ga unstressed array was used for the 70-μm band, while the 160 μm band had a 2 × 20 array of stressed detectors. The detectors are configured in the so-called Z-plane to indicate that the array has substantial size in the third dimension.

An innovative integral field spectrometer, called the Field Imaging Far-Infrared Line Spectrometer (FIFI-LS) was developed for the Herschel Space Observatory and SO-FIA . To accomplish this, the instrument has two 16 × 25 Ge:Ga arrays, unstressed for the 45–110 mm range and stressed for the 110–210 µm range.

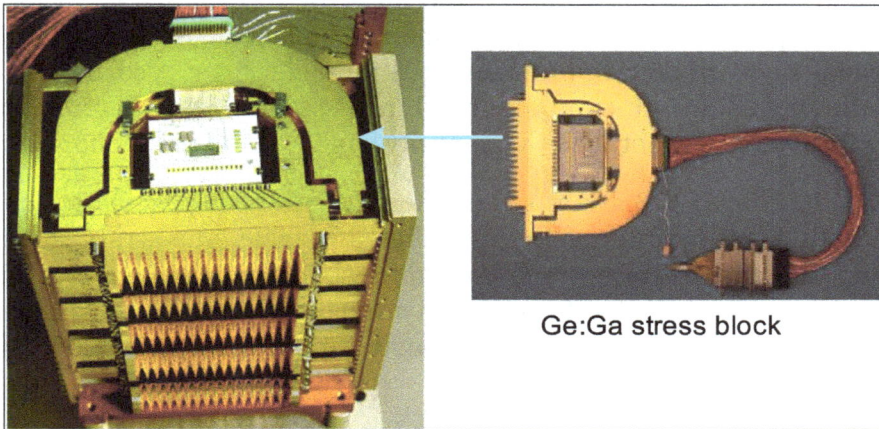

Ge:Ga stress block

PACS photoconductor focal plane array. The 25 stressed and low-stress modules of PACS instrument (corresponding to 25 spatial pixels) in the red and blue arrays are integrated into their housing.

Blocked Impurity Band Detectors

One of the major problems in the design of extrinsic photoconductors is that the doping concentration is driven by conflicting requirements: the doping concentration needs to be as high as possible to get high photon-absorption coefficients (the doping concentration is limited by hopping conduction induced by direct transfer of charge carriers from one impurity site to the next in heavily-doped semiconductors). In contrast a low doping concentration is also desirable to achieve a low electrical conductivity which in turn reduces Johnson noise.

In 1979 M. Petroff and D. Stapelbroeck, working at the Rockwell International Science Centre, invented what they referred to as Blocked-Impurity Band (BIB) detectors. These detectors were developed to provide significantly reduced nuclear radiation sensitivity and improved performance compared to extrinsically doped silicon photoconductors. BIBs have some excellent properties which make them extremely useful for astronomical applications.

BIB detectors overcome the limitation of the doping density present in a standard extrinsic photoconductor by placing a thin intrinsic (undoped) silicon blocking layer between a heavily doped active layer and a planar contact. The active region of detector structure, usually based on epitaxially grown n-type material, is sandwiched between a higher doped degenerate substrate electrode and an undoped blocking layer. Doping of the active layer is high enough for the onset of an impurity band in order to display a high quantum efficiency for impurity ionization (in the case of Si:As BIB, the active layer is doped to $\approx 5 \times 10^{17} cm^{-3}$. The device exhibits a diode-like characteristic, except that photoexcitation of electrons takes place between the donor impurity and the conduction band.

The design of BIB detectors offers a number of advantages over conventional extrinsic photoconductors: the high absorption coefficient of the absorbing layer means that detectors with comparatively small active volumes can be made, providing low susceptibility to cosmic rays without compromising quantum efficiency. Also, due to the heavy doping of the active layer, the impurity band increases in width, therefore effectively decreasing the energy gap between the impurity band and the conduction band.

The main application of BIB arrays today is for ground- and space-based far-infrared astronomy – Si-As BIB performance are gathered in table. The arrays should be operated under the most uniform possible conditions, in the most benign and constant environment possible. Array performance is strongly affected by background levels. Extrinsic silicon arrays for high background applications are less developed than that for low background applications.

The largest extrinsic infrared detector arrays are manufactured for astronomy owing to investments from NASA and the National Science Foundation. At present Raytheon Vision Systems (RVS), DRS Technologies, and Teledyne Imaging Sensors (formerly Rockwell Scientific Company) supply the majority of IR arrays used in astronomy, and between them the most important are BIB detector arrays. Impressive progress has been achieved especially in Si:As BIB array technology with formats as large as 2048 × 2048 and pixels as small as 18 µm.

Table: Performance of Si:As BIB FPAs fabricated in several formats for both ground and space based applications.

Parameter	Si:As BIB	Phenix	MIRI	Aquarius-1k
Application/Users	Ground-based telescopes ESO, Univer. of Tokyo	Space telescopes, JAXA	Space telescopes, JAXA NASA	Ground-based telescopes, ESO, Univer. of Arizona
Format	320 × 240	1024 × 1024, 2048 × 2048	1024 × 1024	1024 × 1024
Pixel size	50 µm	25 µm	25 µm	30 µm
ROIC type	DI	SFD	SFD	SFD
Fill factor	≥95%	≥95%	≥98%	≥98%
ROIC input referred noise	$<1000e^-_{RMS}$	6–$20e^-_{RMS}$	<10–$30e$	Low gain $<1000e^-_{RMS}$ High gain $<100e^-_{RMS}$
Integration capacity	7 or 20×10^6 e$^-$	3×10^6 e$^-$	2×10^5 e$^-$	1 or 11×10^6 e$^-$

Max. frame rates	100–500 Hz	0.1 Hz	0.1 Hz	120 Hz
Number of outputs	16 or 32	4	4	16 or 64
Packaging	LCC	LCC	Module	Module – 2 side buttable

Semiconductor Bolometers

Cooled silicon bolometers demonstrate broadband and nearly flat spectral response in the 1–3000 mm wavelength range. They are easier to fabricate with high operability, good uniformity, and lower cost, but has low operating temperature (4.2–0.3 K). During fabrication of bolometers, their area, operating temperature, thermal time constant, and thermal conductance are adjusted to meet the specific design requirements. The present day technology exists to produce arrays of hundreds of pixels that are operated in many experiments including NASA Pathfinder ground based instruments, and balloon experiments.

The thermistors are typically fabricated by lithography on membranes of Si or SiN. The impedance is selected to a few MΩ to minimize the noise in JFET amplifiers operated at about 100 K. Limitation of this technology is assertion of thermal mechanical, and electrical interface between the bolometers at 100–300 mK and the amplifiers at $\approx$100 K. Usually, JFET amplifiers are sited on membranes which isolate them so effectively that the environment remains at much lower temperatures (about 10 K). In addition, the equipment at 10 K is itself thermally isolated from nearby components at 0.1–0.3 K.

In bolometer metal films that can be continuous or patterned in a mesh absorb the photons. The patterning is designed to select the spectral band, to provide polarization sensitivity, or to control the throughput. Different bolometer architectures are used. In close-packed arrays and spider web, the pop-up structures or two-layer bump bonded structures are fabricated.

At present high-performance bolometer arrays for the far IR and sub-mm spectral ranges are available. For example, the Herschel/PACS instrument uses a 2048-pixel array of bolometers and is an alternative to JFET amplifiers. The architecture of this array is vaguely similar to the direct hybrid mid-infrared arrays, where one silicon wafer is patterned with bolometers, each in the form of a silicon mesh.

Pair Braking Photon Detectors

One of the methods of photon detection consists in using superconducting materials. If the temperature is far below the transition temperature, T_c, most of electrons in them are banded into Cooper pairs. Photons with energies exceeding the binding Cooper pair energies in the superconductor, 2Δ (each electron must be supplied an energy Δ), can break these pairs producing quasiparticles (electrons). When the bias voltage is increased to the gap voltage, the Cooper pairs on one side of the junction can break up into two quasiparticles, which then tunnel to the other side of the junction before recombining, resulting in a sharp increase in current. This process resembles the inter-band absorption in semiconductors, with the energy gap equal 2Δ, when the photons are absorbed and electron-hole pairs are created.

Several structures of pair braking detectors which use different ways to separate quasiparticles from Cooper pairs have been proposed. Among them are: superconductor-insulator-superconductor (SIS) and superconductor-insulator-normal metal (SIN) detectors and mixers, radio frequency (RF) kinetic inductance detectors, and superconducting quantum interference device (SQUID) kinetic inductance detectors. Superconducting detectors offer many benefits: outstanding sensitivity, lithographic fabrication, and large array sizes, especially through the development of multiplexing techniques.

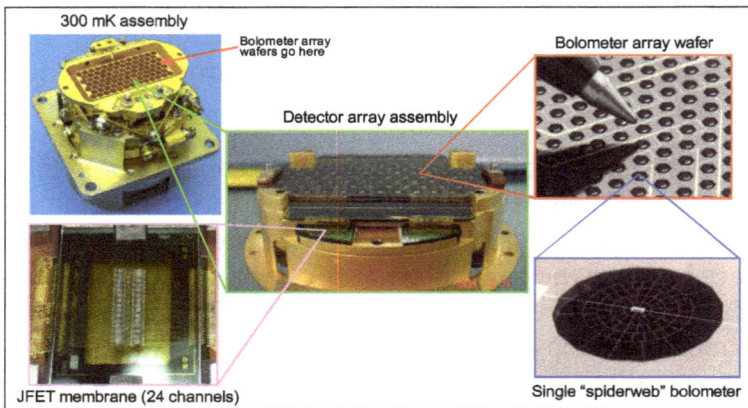

Bolometer array of the Spectral and Photometric Imaging Receiver (SPIRE).

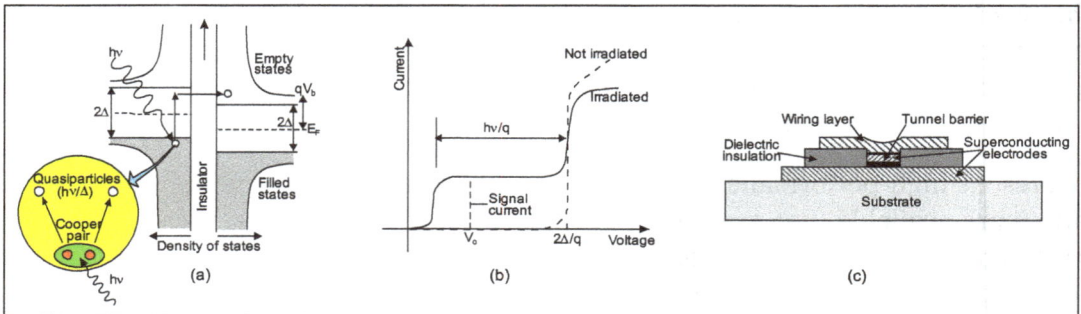

SIS junction: (a) energy diagram with applied bias voltage and illustration of photon assisted tunnelling, (b) current-voltage characteristic of a non-irradiated and irradiated barrier (the intensity of the incident radiation is measured as an excess of the current at a certain bias voltage V_o – schematic creation of quasiparticle is shown inside (a)), and (c) a cross section of a typical SIS junction.

The SIS detector is a sandwich of two superconductors separated by a thin (≈ 20 A) insulator, what is schematically shown in figure Nb and NbTiN are almost exclusively used as superconductors for the electrodes. For a standard junction process, the base electrode is 200-nm sputtered Nb, the tunnel barrier is made using a thin 5-nm sputtered Al layer which is either thermally oxidized (Al_2O_3) or plasma nitridized (AlN). The counterelectrode is 100-nm sputtered Nb or reactively sputtered NbTiN. Typical junction areas are about 1 μm^2.

SIS tunnel junctions are mainly used as mixers in heterodyne type mm and sub-mm receivers, because of their strong nonlinear I–V characteristic. They can be also used as direct detection detectors. Operating temperature of SIS junctions is below 1 K; typically T≤300 mK. However, up to now SIS detectors are difficult to integrate into large arrays.

Progress in development of large-format, high-sensitivity focal plane arrays is especially promising with two detector technologies: transition-edge superconducting (TES) bolometers and microwave kinetic inductance detectors (MKIDs) based on different principles of superconductivity. Multiple instruments are currently in development based on arrays up to 10,000 detectors using both time-domain multiplexing (TDM) and frequency-domain multiplexing (FDM) with superconducting quantum interference devices (SQUIDs). Both sensors show potential to realize the very low ~10^{-20} W/$Hz^{1/2}$ sensitivity needed for space-borne spectroscopy.

A MKID is essentially a high-Q resonant circuit made out of either superconducting microwave transmission lines or a lumped element LC resonator (fabricated from thin aluminium and niobium films). In the first case a meandered quarter-wavelength strip of superconducting material is coupled by means of a coupling capacitance to a coplanar waveguide through-line used for excitation and readout. Lumped element are instead created from an LC series resonant circuit inductively coupled to a microstrip feed line placed in a high frequency resonant circuit. Photons hitting an MKID break Cooper pairs, which changes the surface impedance of the transmission line or inductive element producing a number of quasiparticles. This causes the resonant frequency and quality factor to shift an amount proportional to the energy deposited by the photon. The amplitude (c) and phase (d) of a microwave excitation signal sent through the resonator. The change in the surface impedance of the film following a photon absorption event pushes the resonance to lower frequency and changes its amplitude. The energy of the absorber photon can be determined from the degree of phase and amplitude shift. The readout is almost entirely at room temperature and can be highly multiplexed; in principle hundreds or even thousands of resonators could be read out on a single feedline.

Superconducting HEB and TES Detectors

Among superconducting detectors used for terahertz downconversion, hot-electron-bolometer (HEB) mixers have attracted the attention. Their low local-oscillator (LO) power consumption (less than 1 μW), near-quantum-limited noise performance, and ease of fabrication have placed them above competing technologies in the quest for implementation of large-format heterodyne arrays.

In principle, HEB is quite similar to the transition-edge sensor (TES) bolometer, where small temperature changes caused by the absorption of incident radiation strongly influence resistance of biased sensor near its superconducting transition. The main difference between HEBs and ordinary bolometers is the speed of their response. High speed is achieved by allowing the radiation power to be directly absorbed by the electrons in the superconductor, rather than using a separate radiation absorber and allowing the energy to flow to the superconducting TES via phonons, as ordinary bolometers do. After photon absorption, a single electron initially receives the energy h(n), which is rapidly shared with other electrons, producing a slight increase in the electron temperature. In the next step, the electron temperature subsequently relaxes to the bath temperature through emission of phonons.

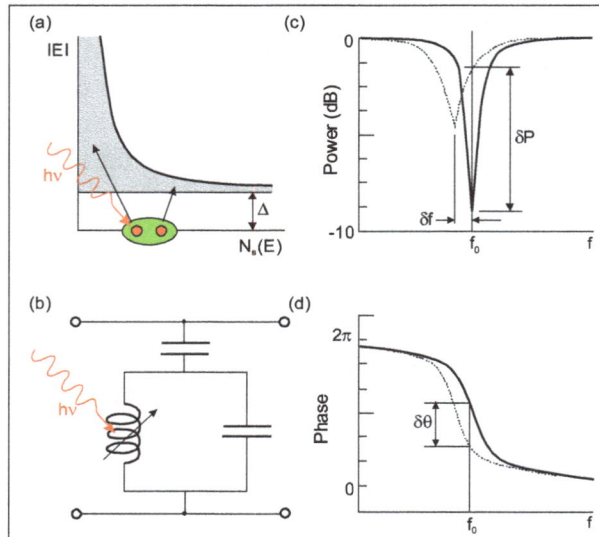

An illustration of the operational principle behind a MKID.

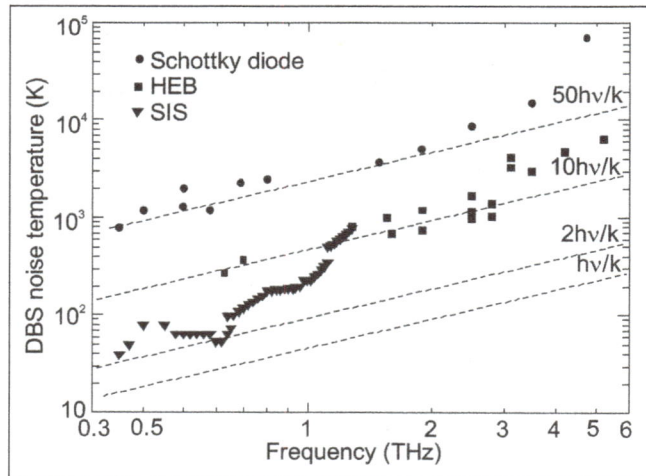

DSB noise temperature of Schottky diode mixers, SIS mixers, and HEB mixers operated in terahertz spectral band.

In comparison with TES, the thermal relaxation time of the HEB's electrons can be made fast by choosing a material with a large electron-phonon interaction. The development of superconducting HEB mixers has lead to the most sensitive systems at frequencies in the terahertz region, where the overall time constant has to be a few tens of picoseconds. These requirements can be realized with a superconducting microbridge made from NbN, NbTiN, or Nb on a dielectric substrate.

Generally, in terahertz receivers, the noise of a mixer is quoted in terms of a single-sideband (SSB), T^{SSB}, or double-sideband (DSB), T^{DSB}, mixer noise temperature. The DSB noise temperatures achieved with Schottky diode mixers, SIS mixers, and HEB mixers operated in terahertz spectral band are presented in figure. The noise temperature of SBD receivers has essentially reached a limit of about 50 hv/k in frequency range below 3 THz. Above 3 THz, there occurs a steep increase, mainly due to increasing losses of the antenna and reduced performance of the diode itself.

The figure presents an example cross-section view of NbN mixer chip. About 150-nm thick Au spiral structure is connected to the contacts pads. The superconducting NbN film extends underneath the contact layer/antenna. The central area of a mixer chip shown in figure. is manufactured from a 3.5-nm thick superconducting NbN film on a high resistive Si substrate. The active NbN film area is determined by the dimensions of the 0.2-μm gap between the gold contact pads. The NbN microstrip is integrated with a planar antenna patterned as log-periodic spiral.

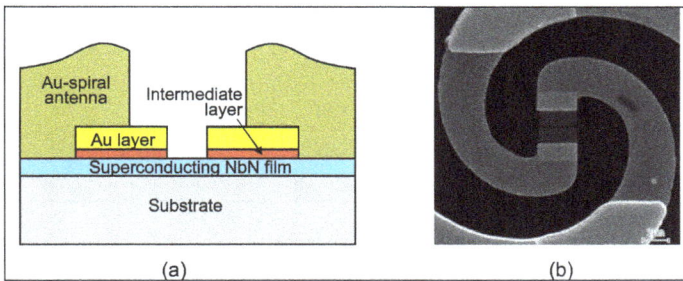

NbN HEB mixer chip: (a) cross section view and (b) SEM micrograph of the central area of mixer.

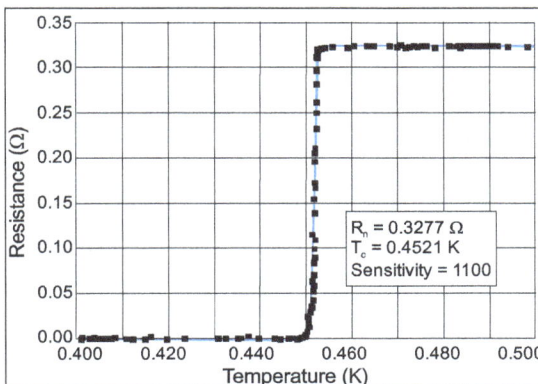

Resistance vs. temperature for a high-sensitivity TES Mo/Au bilayer with superconducting transition at 444 mK.

The name of the transmission-edge sensor (TES) bolometer is derived from its thermometer, which is based on thin superconducting films held within transition region, where it change from the superconducting to the normal state over a temperature range of a few milliKelvin. Changes in temperature transition can be set by using a bilayer film consisting of a normal material and a layer of superconductor (e.g., thin Mo/Au, Mo/Cu, Ti/Au, etc.). Such design enables diffusion of the Cooper pairs from the superconductor into the normal metal and makes it weakly superconducting – this process is called the proximity effect. As a result, the transition temperature is lowered relative to that for the pure superconducting film ($T<200$ mK). Thus in principle, the TES bolometers are quite similar to the HEBs.

TES bolometers are superior to current-biased particle detectors in terms of linearity, resolution, and maximum count rate. At present, theses detectors can be applied for THz photons counting because of their high sensitivity and low thermal time constant. Membrane isolated TES bolometers are capable of reaching a phonon NEP $\approx 4 \times 10^{-20}$ W/Hz$^{1/2}$. The current generation of sub-orbital experiments largely rely on TES bolometers. Important feature of this sensor is that it can operate in wide spectral band, between the radio and gamma rays.

The most ambitious example of TES bolometer array is that used in the submillimetre camera SCUBA (Submillimetre Common-User Bolometer Array) 2 with 10,240 pixels. The camera operated at wavelengths of 450 and 850 μm has been mounted on the James Clerk Maxwell Telescope in Hawaii. Each SCUBA-2 array is made of four side-buttable sub-arrays, each with 1280 (32×40) transition-edge sensors.

BOLOMETERS

A bolometer is a device for measuring the power of incident electromagnetic radiation via the heating of a material with a temperature-dependent electrical resistance. It was invented in 1878 by the American astronomer Samuel Pierpont Langley.

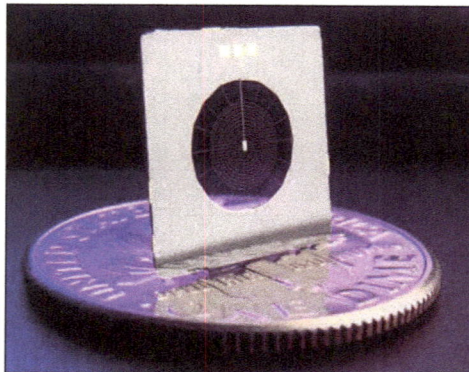

Spiderweb bolometer for measurements of the cosmic microwave background radiation.

Principle of Operation

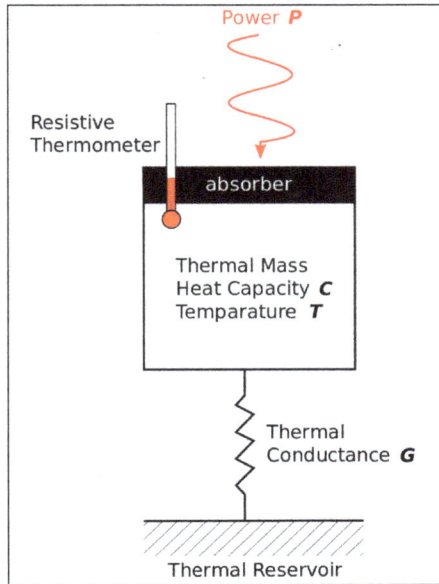

Conceptual schematic of a bolometer. Power, P, from an incident signal is absorbed and heats up a thermal mass with heat capacity, C, and temperature, T. The thermal mass is connected to a reservoir of constant temperature through a link with thermal conductance, G. The temperature increase is $\Delta T = P/G$ and is measured with a resistive thermometer, allowing the determination of P. The intrinsic thermal time constant is $\tau = C/G$.

A bolometer consists of an absorptive element, such as a thin layer of metal, connected to a thermal reservoir (a body of constant temperature) through a thermal link. The result is that any radiation impinging on the absorptive element raises its temperature above that of the reservoir – the greater the absorbed power, the higher the temperature. The intrinsic thermal time constant, which sets the speed of the detector, is equal to the ratio of the heat capacity of the absorptive element to the thermal conductance between the absorptive element and the reservoir. The temperature change can be measured directly with an attached resistive thermometer, or the resistance of the absorptive element itself can be used as a thermometer. Metal bolometers usually work without cooling. They are produced from thin foils or metal films. Today, most bolometers use semiconductor or superconductor absorptive elements rather than metals. These devices can be operated at cryogenic temperatures, enabling significantly greater sensitivity.

Bolometers are directly sensitive to the energy left inside the absorber. For this reason they can be used not only for ionizing particles and photons, but also for non-ionizing particles, any sort of radiation, and even to search for unknown forms of mass or energy (like dark matter); this lack of discrimination can also be a shortcoming. The most sensitive bolometers are very slow to reset (i.e., return to thermal equilibrium with the environment). On the other hand, compared to more conventional particle detectors, they are extremely efficient in energy resolution and in sensitivity. They are also known as thermal detectors.

Langley's Bolometer

The first bolometers made by Langley consisted of two steel, platinum, or palladium foil strips covered with lampblack. One strip was shielded from radiation and one exposed to it. The strips formed two branches of a Wheatstone bridge which was fitted with a sensitive galvanometer and connected to a battery. Electromagnetic radiation falling on the exposed strip would heat it and change its resistance. By 1880, Langley's bolometer was refined enough to detect thermal radiation from a cow a quarter of a mile away. This radiant-heat detector is sensitive to differences in temperature of one hundred-thousandth of a degree Celsius (0.00001 C). This instrument enabled him to thermally detect across a broad spectrum, noting all the chief Fraunhofer lines. He also discovered new atomic and molecular absorption lines in the invisible infrared portion of the electromagnetic spectrum. Nikola Tesla personally asked Dr. Langley if he could use his bolometer for his power transmission experiments in 1892. Thanks to that first use, he succeeded in making the first demonstration between West Point and his laboratory on Houston Street.

Applications in Astronomy

While bolometers can be used to measure radiation of any frequency, for most wavelength ranges there are other methods of detection that are more sensitive. For sub-millimeter wavelengths (from around 200 μm to 1 mm wavelength, also known as the far-infrared or terahertz), bolometers are among the most sensitive available detectors, and are therefore used for astronomy at these wavelengths. To achieve the best sensitivity, they must be cooled to a fraction of a degree above absolute zero (typically from 50 mK to 300 mK). Notable examples of bolometers employed in submillimeter astronomy include the Herschel Space Observatory, the James Clerk Maxwell Telescope, and the Stratospheric Observatory for Infrared Astronomy (SOFIA).

Applications in Particle Physics

The term bolometer is also used in particle physics to designate an unconventional particle detector. They use the same principle described above. The bolometers are sensitive not only to light but to every form of energy. The operating principle is similar to that of a calorimeter in thermodynamics. However, the approximations, ultra low temperature, and the different purpose of the device make the operational use rather different. In the jargon of high energy physics, these devices are not called "calorimeters", since this term is already used for a different type of detector. Their use as particle detectors was proposed from the beginning of the 20th century, but the first regular, though pioneering, use was only in the 1980s because of the difficulty associated with cooling and operating a system at cryogenic temperature. They can still be considered to be at the developmental stage.

Microbolometers

A microbolometer is a specific type of bolometer used as a detector in a thermal camera. It is a grid of vanadium oxide or amorphous silicon heat sensors atop a corresponding grid of silicon. Infrared radiation from a specific range of wavelengths strikes the vanadium oxide or amorphous silicon, and changes its electrical resistance. This resistance change is measured and processed into temperatures which can be represented graphically. The microbolometer grid is commonly found in three sizes, a 640×480 array, a 320×240 array (384×288 amorphous silicon) or less expensive 160×120 array. Different arrays provide the same resolution with larger array providing a wider field of view. Larger, 1024×768 arrays were announced in 2008.

Hot Electron Bolometer

The hot electron bolometer (HEB) operates at cryogenic temperatures, typically within a few degrees of absolute zero. At these very low temperatures, the electron system in a metal is weakly coupled to the phonon system. Power coupled to the electron system drives it out of thermal equilibrium with the phonon system, creating hot electrons. Phonons in the metal are typically well-coupled to substrate phonons and act as a thermal reservoir. In describing the performance of the HEB, the relevant heat capacity is the electronic heat capacity and the relevant thermal conductance is the electron-phonon thermal conductance.

If the resistance of the absorbing element depends on the electron temperature, then the resistance can be used as a thermometer of the electron system. This is the case for both semiconducting and superconducting materials at low temperature. If the absorbing element does not have a temperature-dependent resistance, as is typical of normal (non-superconducting) metals at very low temperature, then an attached resistive thermometer can be used to measure the electron temperature.

Microwave Measurement

A bolometer can be used to measure power at microwave frequencies. In this application, a resistive element is exposed to microwave power. A dc bias current is applied to the resistor to raise its temperature via Joule heating, such that the resistance is matched to the waveguide characteristic impedance. After applying microwave power, the bias current is reduced to return the bolometer to its resistance in the absence of microwave power. The change in the dc power is then equal to the absorbed microwave power. To reject the effect of ambient temperature changes, the active (measuring) element is in a bridge circuit with an identical element not exposed to microwaves;

variations in temperature common to both elements do not affect the accuracy of the reading. The average response time of the bolometer allows convenient measurement of the power of a pulsed source.

MICROBOLOMETER

A microbolometer is a specific type of bolometer used as a detector in a thermal camera. Infrared radiation with wavelengths between 7.5–14 μm strikes the detector material, heating it, and thus changing its electrical resistance. This resistance change is measured and processed into temperatures which can be used to create an image. Unlike other types of infrared detecting equipment, microbolometers do not require cooling.

Theory of Construction

A microbolometer is an uncooled thermal sensor. Previous high resolution thermal sensors required exotic and expensive cooling methods including stirling cycle coolers and liquid nitrogen coolers. These methods of cooling made early thermal imagers expensive to operate and unwieldy to move. Also, older thermal imagers required a cool down time in excess of 10 minutes before being usable.

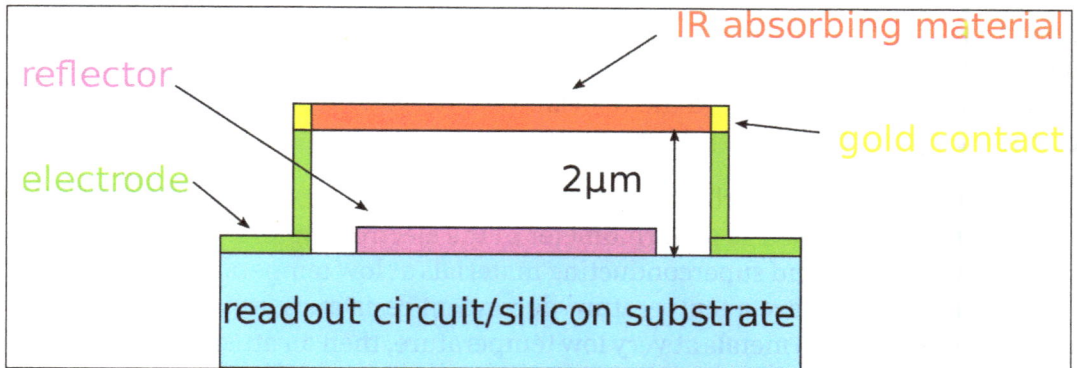

Cross-sectional view of a microbolometer.

A microbolometer consists of an array of pixels, each pixel being made up of several layers. The cross-sectional diagram shown in figure provides a generalized view of the pixel. Each company that manufactures microbolometers has their own unique procedure for producing them and they even use a variety of different absorbing materials. In this example the bottom layer consists of a silicon substrate and a readout integrated circuit (ROIC). Electrical contacts are deposited and then selectively etched away. A reflector, for example, a titanium mirror, is created beneath the IR absorbing material. Since some light is able to pass through the absorbing layer, the reflector redirects this light back up to ensure the greatest possible absorption, hence allowing a stronger signal to be produced. Next, a sacrificial layer is deposited so that later in the process a gap can be created to thermally isolate the IR absorbing material from the ROIC. A layer of absorbing material is then deposited and selectively etched so that the final contacts can be created. To create the final bridge like structure shown in figure, the sacrificial layer is removed so that the absorbing material is suspended approximately 2 μm above the readout circuit. Because microbolometers do not undergo any cooling, the absorbing material must be thermally isolated from the bottom ROIC and the bridge like structure allows for this to occur. After the array of pixels is created the microbolometer is encapsulated under a vacuum to increase the longevity of the device. In some cases the entire fabrication process is done without breaking vacuum.

The quality of images created from microbolometers has continued to increase. The microbolometer array is commonly found in two sizes, 320×240 pixels or less expensive 160×120 pixels. Current technology has led to the production of devices with 640×480 or 1024x768 pixels. There has also been a decrease in the individual pixel dimensions. The pixel size was typically 45 μm in older devices and has been decreased to 12 μm in current devices. As the pixel size is decreased and the number of pixels per unit area is increased proportionally, an image with higher resolution is created, but with a higher NETD (Noise Equivalent Temperature Difference (differential)) due to smaller pixels being less sensitive to IR radiation.

Detecting Material Properties

There is a wide variety of materials that are used for the detector element in microbolometers. A main factor in dictating how well the device will work is the device's responsivity. Responsivity is the ability of the device to convert the incoming radiation into an electrical signal. Detector material properties influence this value and thus several main material properties should be investigated: TCR, 1/f Noise, and Resistance.

Temperature Coefficient of Resistance (TCR)

The material used in the detector must demonstrate large changes in resistance as a result of minute changes in temperature. As the material is heated, due to the incoming infrared radiation, the resistance of the material decreases. This is related to the material's temperature coefficient of resistance (TCR) specifically its negative temperature coefficient. Industry currently manufactures microbolometers that contain materials with TCRs near −2 %/K. Although many materials exist that have far higher TCRs, there are several other factors that need to be taken into consideration when producing optimized microbolometers.

1/f Noise

1/f noise, like other noises, causes a disturbance that affects the signal and that may distort the information carried by the signal. Changes in temperature across the absorbing material are determined by changes in the bias current or voltage flowing through the detecting material. If the noise is large then small changes that occur may not be seen clearly and the device is useless. Using a detector material that has a minimum amount of 1/f noise allows for a clearer signal to be maintained between IR detection and the output that is displayed. Detector material must be tested to assure that this noise does not significantly interfere with signal.

Resistance

Using a material that has low room temperature resistance is also important. Lower resistance across the detecting material mean less power will need to be used. Also, there is a relationship between resistance and noise, the higher the resistance the higher the noise. Thus, for easier detection and to satisfy the low noise requirement, resistance should be low.

Detecting Materials

The two most commonly used IR radiation detecting materials in microbolometers are amorphous silicon and vanadium oxide. Much research has been done to test the feasibility of other materials to be used. Those investigated include: Ti, YBaCuO, GeSiO, poly SiGe, BiLaSrMnO and protein-based cytochrome C and bovine serum albumin.

Amorphous Si (a-Si) works well because it can easily be integrated into the CMOS fabrication process, is highly stable, a fast time constant, and has a long mean time before failure. To create the layered structure and patterning, the CMOS fabrication process can be used but it requires temperatures to stay below 200°C on average. A problem with some potential materials is that to create the desirable properties their deposition temperatures may be too high although this is not a problem for a-Si thin films. a-Si also possesses excellent values for TCR, 1/f noise and resistance when the deposition parameters are optimized.

Vanadium oxide thin films may also be integrated into the CMOS fabrication process although not as easily as a-Si for temperature reasons. VO is an older technology than a-Si, and for these reasons its performance and longevity are less. Deposition at high temperatures and performing post-annealing allows for the production of films with superior properties although acceptable films can still be made subsequently fulfilling the temperature requirements. VO_2 has low resistance but undergoes a metal-insulator phase change near 67 °C and also has a lower value of TCR. On the other hand, V_2O_5 exhibits high resistance and also high TCR. Many phases of VO_x exist although it seems that $x \approx 1.8$ has become the most popular for microbolometer applications.

Active vs. Passive Microbolometers

Most microbolometers contain a temperature sensitive resistor which makes them a passive electronic device. In 1994 one company, Electro-Optic Sensor Design (EOSD), began looking into producing microbolometers that used a thin film transistor (TFT), which is a special kind of field effect transistor. The main change in these devices would be the addition of a gate electrode. Although the main concepts of the devices are similar, using this design allows for the advantages of the TFT to be utilized. Some benefits include tuning of the resistance and activation energy and the reduction of periodic noise patterns. As of 2004 this device was still being tested and was not used in commercial IR imaging.

Advantages

- They are small and lightweight. For applications requiring relatively short ranges, the physical dimensions of the camera are even smaller. This property enables, for example, the mounting of uncooled microbolometer thermal imagers on helmets.

- Provide real video output immediately after power on.

- Low power consumption relative to cooled detector thermal imagers.

- Very long MTBF.

- Less expensive compared to cameras based on cooled detectors.

Disadvantages

- Less sensitive (due to higher noise) than cooled thermal and photon detector imagers, and as a result have not been able to match the resolution of cooled semiconductor based approaches.

Performance Limits

The sensitivity is partly limited by the thermal conductance of the pixel. The speed of response is limited by the thermal heat capacity divided by the thermal conductance. Reducing the heat capacity increases the speed but also increases statistical mechanical thermal temperature fluctuations (noise). Increasing the thermal conductance raises the speed, but decreases sensitivity.

Origins

Microbolometer technology was originally developed by Honeywell starting in the late 1970s as a classified contract for the US Department of Defense. The US Government declassified the technology in 1992. After declassification Honeywell licensed their technology to several manufacturers.

Golay Cell

The Golay cell is a type of opto-acoustic detector mainly used for infrared spectroscopy. It consists of a gas-filled enclosure with an infrared absorbing material and a flexible diaphragm or membrane. When infrared radiation is absorbed, it heats the gas, causing it to expand. The resulting increase in pressure deforms the membrane. Light reflected off the membrane is detected by a photodiode, and motion of the membrane produces a change in the signal on the photodiode. The concept was originally described in 1947 by Marcel J. E. Golay, after whom it came to be named.

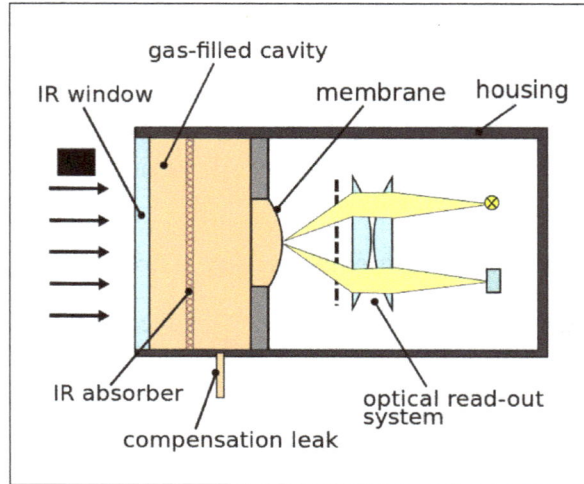

Schematic of a Golay cell.

The Golay cell has high sensitivity and a flat response over a very broad range of frequencies. The response time is modest, of order 10 ms. The detector performance is degraded in the presence of mechanical vibrations.

PYROELECTRIC DETECTORS

Pyroelectric detectors are sensors for light which are based on the pyroelectric effect. They are widely used for detecting laser pulses (rather than continuous-wave light), often in the infrared spectral region, and with the potential for a very broad spectral response. Pyroelectric detectors are used as the central parts of many optical energy meters, and are typically operated at room temperature (i.e., not cooled). Compared with energy meters based on photodiodes, they can have a much broader spectral response.

There are various other applications of pyroelectric sensors, for example fire detection, satellite-based infrared detection, and the detection of persons via their infrared emission (*motion detectors*).

Operation Principle

We first consider the basic operation principle. A pyroelectric detector contains a piece of ferroelectric crystal material with electrodes on two sides – essentially a capacitor.

One of those electrodes has a black coating (or a processed absorbing metal surface), which is exposed to the incident radiation. The incident light is absorbed on the coating and thus also causes some heating of the crystal, because the heat is conducted through the electrode into the crystal. As a result, the crystal produces some pyroelectric voltage; one can electronically detect that voltage or alternatively the current when the voltage is held constant. For a constant optical power, that pyroelectric signal would eventually fade away; the device would therefore not be suitable for measuring the intensity of continuous-wave radiation. Instead, such a detector is usually used with light pulses; in that case, one obtains a bipolar pulse structure, where one initially obtains a voltage in one direction and after the pulse a voltage in the opposite direction.

Pyroelectric detectors.

Due to that operation principle, pyroelectric detectors belong to the thermal detectors: they do not directly respond to radiation, but only to the generated heat.

In the simple explained form, the detector would be relatively sensitive to fluctuations of the ambient temperature. Therefore, one often uses an additional compensating crystal, which is exposed to essentially the same temperature fluctuations but not to the incoming light. By taking the difference of signals from both crystals, one can effectively reduce the sensitivity to external temperature changes.

The pyroelectric charges are typically detected with an operational amplifier (OpAmp) based on field-effect transistors (JFETs) with very low leakage current.

Ferroelectric Crystal Materials

Only a small group of crystals possesses a low enough crystal symmetry (e.g. monoclinic) for exhibiting ferroelectric properties and the pyroelectric effect. They have an electrical polarization which is temperature-dependent and thus leads to pyroelectric charges when the temperature changes.

A particularly high sensitivity is achieved when using triglycine sulfate (TGS, $(NH_2CH_2COOH)_3 \cdot H_2SO_4$). That material, however, has a rather low Curie temperature of 49 °C;

above that temperature, the ferroelectric properties vanish. A somewhat higher Curie temperature of 61 °C is obtained for modified form of that material, deuterated triglycine sulfate (DTGS). Both materials, however, are not acceptable for applications where one cannot ensure that one always stays sufficiently below the Curie temperature. Note also that the pyroelectric response is substantially increased just below the Curie temperature, so that the calibration is affected. Further, there is a risk of depoling at higher temperatures. In addition, TGS and DTGS are water-solvable, hygroscopic and fragile, therefore not well suited for robust optical energy meters.

Other ferroelectric materials, belonging to the perovskite group, are lead zirconate titanate (PZT, $PbZrTiO_3$) and lead titanate (PT, $PbTiO_3$). They need to be used in a ceramic form (e.g. as deposited thin films), since large crystals are hard to make; additional dopants are required for stability at room temperature. These materials can be produced at relatively low cost and are far more robust than TGS.

A material with very high Curie temperature and overall high robustness is lithium tantalate ($LiTaO_3$), which is therefore often used despite its lower pyroelectric response.

Performance Parameters

Spectral Response

As usual for thermal detectors, the spectral response can be very broad; one only requires sufficiently broadband absorption.

A pyroelectric sensor may be equipped with an infrared filter which transmits only light in a certain range of wavelengths.

Active Area

The active area is usually a circular disk or a rectangular area with a diameter between a few millimeters and a few tens of millimeters. Tentatively, detectors for higher pulse energies have larger active areas.

Surface Reflectivity

In principle, a pyroelectric detector should ideally absorb all incident light for having a sensitivity as high as possible. However, for a fast response one wants to use a thin absorbing coating, which sits on a reflecting metallic electrode, or just a metallic electrode with processed surface structure for enhanced absorption. Therefore, there can be a substantial reflectivity (of the order of 50%) in practice.

Maximum Pulse Width

For such a detector to work properly, the input pulses need to be sufficiently short. The maximum allowed pulse width vary substantially between different models; it can be some tens of microseconds, for example. Pulses from a Q-switched laser are always short enough.

Sensitivity and Dynamic Range

Pyroelectric detectors are normally used for detecting pulses with pulse energies in the nanojoule or microjoule region. The most sensitive devices have a noise floor well below 100 pJ, so that even pulse energies of a few nanojoules can be measured with a reasonable accuracy. At the same time, pulse energies as high as 10 µJ may be allowed, so that one effectively has a dynamic range of e.g. 40 dB for energy measurements.

Other devices are optimized for much higher pulse energies of e.g. several joules, but have a higher noise floor, allowing measurements down to pulse energies of tens of microjoules instead of nanojoules.

Note that there may be a further limitation for the allowed average power. That means that for the highest possible pulse repetition rates, the pulse energy needs to be limited, because otherwise there would be too strong heating of the sensor.

Detection Bandwidth

A typical detection bandwidth of a pyroelectric detector is several kilohertz, or sometimes even tens of kilohertz. This is quite fast compared with many other thermal detectors such as thermocouples and thermopiles, and is possible due to the small thermal capacity of the compact detector crystal. (The electrical capacity can in principle also be a limiting factor, but typically the thermal relaxation time is essential.) For a particularly fast response, one can use thin metallic electrodes with a processed absorbing surface, minimizing the thermal capacity.

Suppliers often specify instead of a true bandwidth the maximum allowed pulse repetition rate, where one can still measure the energy of each pulse. This is actually the quantity which is most relevant for users. One can use such a detector for monitoring pulse energy fluctuations of a Q-switched laser, for example. For measuring only the average pulse energy, one could simply use a slow thermal detector, which delivers the average power, and divide this by the pulse repetition rate.

The pulse repetition rate of a mode-locked laser would be far too high; here one would have to use a photodiode.

Response to Sound (Microphony)

All pyroelectric crystal materials are also piezoelectric. Therefore, a pyroelectric detective will also show some response to incoming sound waves, i.e., it acts as a microphone – which is normally unwanted. Such microphony can be suppressed e.g. with proper mounting and shielding of the crystal.

References

- Terahertz-detectors- 319172386: researchgate.net, Retrieved 28 January, 2019

- Wang, Hongchen; Xinjian Yi; Jianjun Lai & Yi Li (31 January 2005). "Fabricating Microbolometer Array on Unplanar Readout Integrated Circuit". International Journal of Infrared and Millimeter Waves. 26 (5): 751–762. Bibcode:2005IJIMW..26..751W. doi:10.1007/s10762-005-4983-8

- D. Klocke, A. Schmitz, H. Soltner, H. Bousack and H. Schmitz, "Infrared receptors in pyrophilous ('fire loving') insects as model for new un-cooled infrared sensors," Beilstein Journal of Nano-technology 2, 186 (2011), doi:10.3762/bjnano.2.22

- Pyroelectric-detectors: rp-photonics.com, Retrieved 06 July, 2019

8

Organic Photodetector

Organic photodetector consists of metal electrodes and organic small molecules as donors with fullerene derivatives as acceptors. Organic semiconductor and organic photodetector fabrication are some of its elements. This chapter has been carefully written to provide an easy understanding of organic photodetectors.

Photodetectors are sensors of electromagnetic radiation (i.e. photons) that convert the optical signal into an electric signal proportional to the radiation intensity. The capability of certain materials to generate current when exposed to light was noticed first by Antoine Becquerel in 1850 but the first theoretical explanation of this phenomenon was given in 1935 by Vladimir Zworykin, George Ashmun Morton, and Louis Malter of RCA, where three kinds of light detection were identified: potential barrier crossing (used in Schottky photodiodes), linked state-free state transition (related to quantum well detectors) and electron-hole pair generation, mechanism that governs the photodetection in photoconductors and photodiodes.

Working Principle

The device structure consisted of an organic semiconductor film sandwiched between two metal contacts as shown in the figure. A photodiode can operate in photovoltaic or photoconductive mode. In photoconductive mode, the photodiode is operated under the reverse bias and subjected to the illumination, providing a photoconductive current. In photovoltaic mode, a bias is not provided but the device is given irradiation to supply an output voltage to drive current through an external circuit.

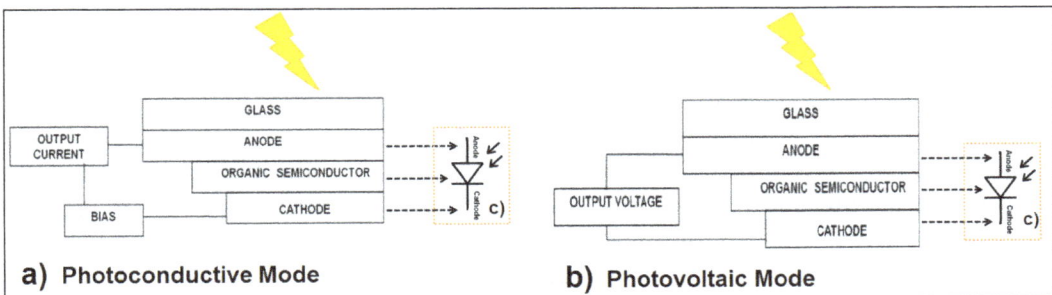

Representation of a a)photoconductive and b) photovoltaic mode of a photodiode, and c) the ideal electrical symbol of a photodiode.

Photocurrent Generation in Organic Semiconductors

The photocurrent in organic devices is generated through combination of five processes as follows: photon absorption, exciton formation, exciton diffusion, exciton dissociation and finally, charge transport and collection.

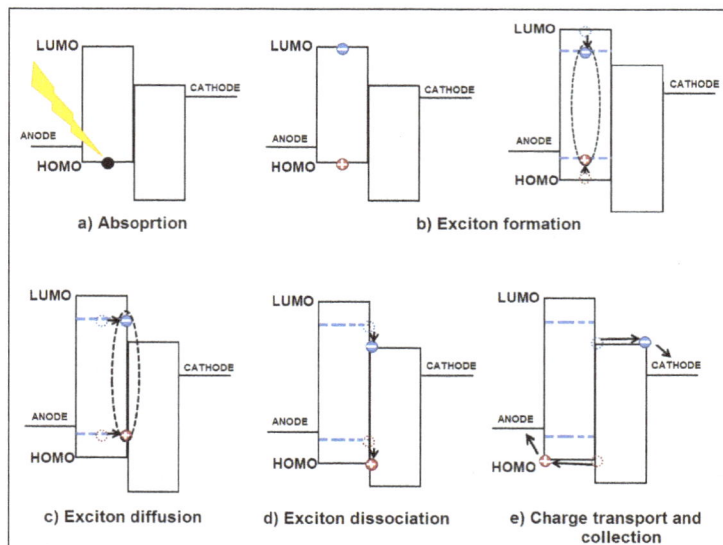

Photodetection process in an organic donor/acceptor system.

The major contributor in the mechanisms of photocurrent generation is the quasi-particle exciton, an electron-hole pair bound by Coulomb forces. This is the main differentiator in the photodetection process between inorganic and organic semiconductors.

The process starts with the absorption of the photons if this condition is fulfilled: the energy of the photon must be greater than the semiconductor band gap. Therefore, an electron can be excited from a low to a high energy state. For organics, the energy band gap is defined as the difference between the highest occupied molecular orbital (HOMO) and the lowest unoccupied molecular orbital (LUMO).

Once the photon energy is absorbed, it elevates the energy of the carrier from a ground state to a photo-induced excited state (blue dashed line in figure). Absorption of a photon by excitation of an electron from the HOMO state to an excited state above the LUMO level leaves behind an unoccupied valence state, termed as hole, and the photon energy is the potential energy difference between this excited electron-hole pair.

At this stage, the exciton migrates to the junction. This is the step where the process in an organic semiconductor differs from the one in inorganic semiconductors: excitons must be formed and disassociated to generate a current. Organic devices have two components: an electron donor and an electron acceptor with a corresponding interface between them, where an exciton must reach during its lifetime in order to disassociate. Then, the bound exciton is dissociated into an electron and hole. This is done by introducing a second organic semiconductor, with a lower LUMO level, such that electron transfer between the two types of semiconductor is favorable. That is the reason for the material with the highest LUMO to be called electron donor, while the other is called the electron acceptor.

For the electron to be transferred, the exciton bond energy must be lower than the difference between LUMO of the donor and the LUMO of the acceptor. This condition is valid when the excitons are generated in the donor phase and electrons are transferred to the acceptor. On the contrary, (excitons are generated in the acceptor phase), the following condition must be fulfilled in order for holes to be transferred from the acceptor to the donor: the exciton bond energy must be lower than the difference between HOMO of the donor and the HOMO of the acceptor.

As a last step to generate current, the separated charges (electron and hole) will travel in the acceptor and donor phases, respectively, in order to be collected by each electrode. The current flow is controlled by proper selection of the electrodes having different work functions. In that sense, the anode electrode is chosen as a high work function material and the cathode selected as a low work function metal, so that holes will go to the high work function anode and the electrons to the low work function cathode. Finally, the interaction of current with external circuit will provide the output signal.

Planar and Bulk Heterojunction Photodetectors

During absorption, photons need to pass through a certain thickness of the organic semiconductor layer (i.e. active layer of the device) approximately on the order of 100 nm, in order to be absorbed. At the same time, this active layer should be thin enough, in order to ensure exciton dissociation at the donor-acceptor interface, because charge carriers in organic semiconductors have diffusion lengths of only 3 to 10 nm. Hence, the donor-acceptor structure in the active layer becomes a critical feature to be consider for efficient performance of organic devices.

There are two different structures for donor-acceptor heterojunctions: planar and bulk heterojunctions.

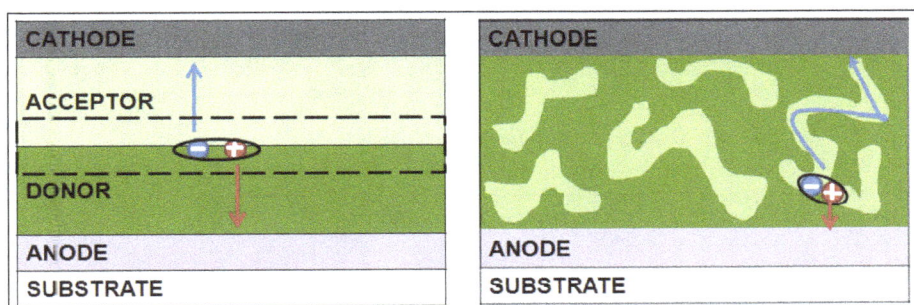

Donor/Acceptor a) Planar Heterojunction and b) Bulk Heterojunction.

A planar heterojunction refers to an organic film formed by superposition of an acceptor on top of a donor layer. As the exciton must travel up to the donor-acceptor interface to dissociate, the thickness of a planar structure must be in the range of the diffusion lenght ($\approx$ 10 nm) to avoid recombination before dissociation. Moreover, just a region in the order of 10 nm, on each side of the junction would contribute to current generation, whereas around 100 nm of material is needed for absorption. The optimal structure is obtained when the junction area is increased while keeping the adequate thickness for a maximum light absorption. This is the case of the bulk heterojunction, a 3D nanoscale phase separation of donor and acceptor upon drying, which effectively distribute the heterojunction throughout the bulk of the active layer. Here, the morphology of the layer must be controlled so that the electron and the hole can always go back to the cathode and anode respectively through the percolated paths all over the bulk and avoid their recombination. One method to fabricate this active layer is by dissolving and mixing the donor and acceptor materials in a common solvent, thus, the active layer is formed from the solution mixture. This also allows the fabrication by simple processes such as spin coating.

Another concept regarding the architecture of the device is now introduced: the device geometry. There are two common types in OPD: standard and inverted. The difference between both of them resides in the nature of electrodes. When the top and bottom electrodes are the cathode and the anode respectively, the architecture follows a standard geometry, in which the electrons are collected in the top electrode and the hole in the bottom contact. For the inverted geometry, the polarity is the opposite. The anode in this case is in the top and the cathode in the bottom, so that the charge transport and collection is opposite to the standard structure (electrons to the bottom contact and holes to the top contact, as illustrated in figure). The performance of devices from both geometries are comparable, but the inverted architecture offers more air-stability because the high work function metal contact is on top instead of in the bottom as for the case of the standard geometry.

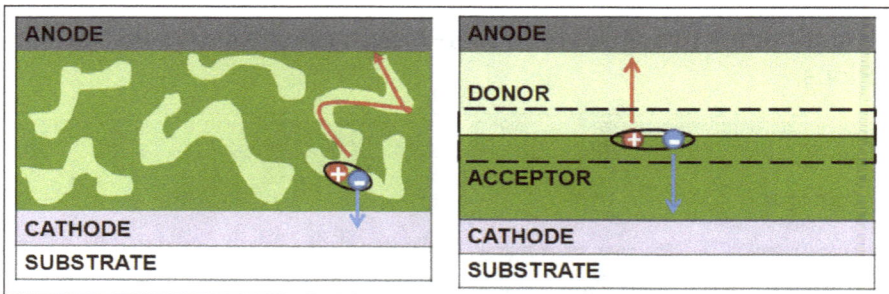

Inverted geometry.

Current-Voltage Characteristics

The traditional view to model the current in donor-acceptor organic devices is focused on the Shockley diode equation modified for a series and shunt resistance, as expressed by the equation:

$$J = \frac{R_{sh}}{R_s + R_{sh}} \left\{ J_0 \left[\exp\left(\frac{q(V - JR_s)}{nkT} \right) - 1 \right] + \frac{V}{R_{sh}} \right\}$$

Where J is the current density, R_s is the series resistance, R_{sh} is the shunt resistance, J_0 is the dark saturation current density, n is the ideality factor and V is the applied bias voltage.

Once the equation is plotted, the characteristic I-V curve is obtained under dark and illumination conditions as shown in the figure. The regime when the applied voltage is greater than zero $(V_{bias} > 0)$, is called forward bias, here the current increases exponentially, because the polarity conditions of the electrodes is favorable for the generation of photocurrent, as the anode becomes more positive and the cathode more negative. On the contrary, under reverse bias $(V_{bias} < 0)$, the measured current is saturated and it is extremely low, because the anode becomes more negative and the cathode more positive, creating an energy barrier for the carriers to be injected from the electrodes to the organic film.

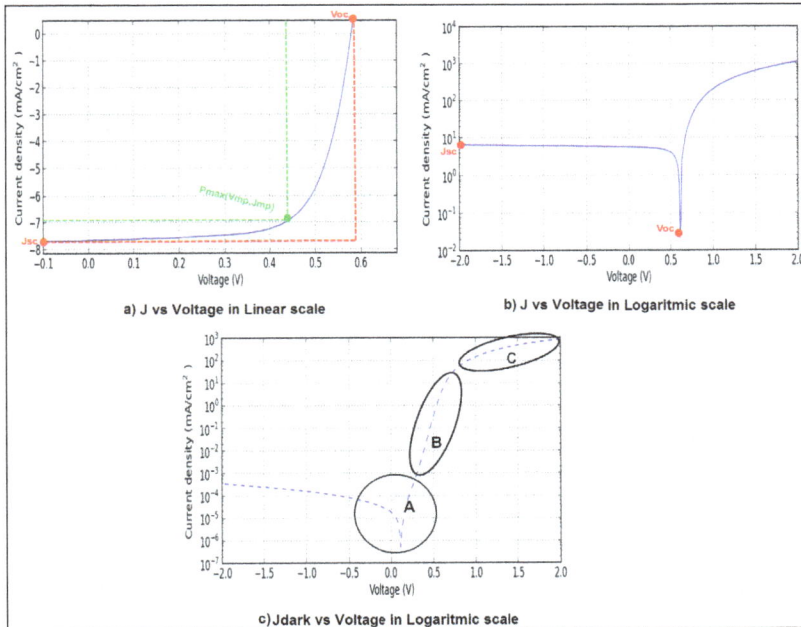

Characteristic JV curve of a photodiode.

Different regimes can be identified from the IV curve under dark conditions, in logaritmic scale:

- A: A linear regime at negative voltages and low positive voltages where the current is limited mainly by the shunt resistance.

- B: An exponential behavior at intermediate positive voltages where the current is controlled by the diode.

- C: A second linear regime at high voltages where the current is limited by the series resistance.

Finally, some parameters can be extracted and calculated from the IV curve, when the photodiode is operating under illumination:

- Open circuit voltage (Voc): The photocurrent in reverse bias, the electrons flow toward the cathode and the holes flow to the anode. When a forward voltage is applied, it compensates that reverse photocurrent until a point where the current is zero. This point is known as open circuit voltage, because even with an applied voltage different to zero, there is no current, so the electrical circuit of the system is open.

- Short circuit current (Isc): When there is a current without any external applied voltage, the current is referred to as the short circuit photocurrent. For instance, the Jsc (short circuit current density) seen in the figure is approximately $8 \, mA / cm^2$.

- Fill factor (FF): It is a measure of the "squareness" through the area of the largest rectangle which fits in the IV curve (in the region before the Voc is reached). It can be estimated, once the points that form the maximum square fitting the IV curve are identified. This points are called the voltage at maximum power (Vmp) and the current (or current density) at maximum power (Jmp). In the case of the device whose IV curve is plotted in figure, the Vmp is 0.45 V and the Jmp is $7 \, mA / cm^2$. Finally, the FF is calculated from the Equation 2.2, resulting in a FF of 70% for this example.

$$FF = \frac{JmpVmp}{JscVoc}$$

- Power conversion efficiency (η): The efficiency is determined as the fraction of incident power (Pin) that is converted to electricity. It is calculated as shown by the equation 2.3. For the P3HT:PCBM system of this example, the efficiency is 3.5%.

$$\eta = \frac{IscVocFF}{Pin}$$

These parameters are usually determined to characterize a solar cell. Last but not least, two more parameters can be extracted from the IV curve: parasitic resistances: series Rs and shunt (or parallel) Rsh. Ideally, the photodiode is considered as illustrated in the figure, but a more realistic electrical representation of the photodiode is as shown in figure. Here, the Rs decreases the current density as it increases and the Rsh causes power losses and decreases the open circuit voltage by generating an alternate current path for the photocurrent. Therefore, Rs must be smaller than tens of Ω and Rsh must be greater than tens of MΩ.

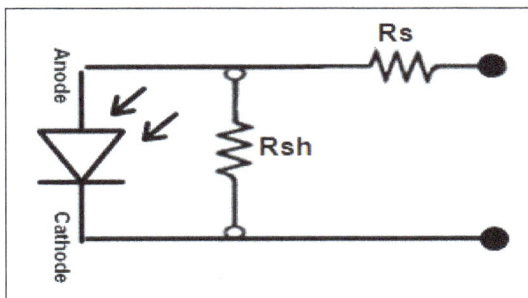

Photodiode realistic electrical circuit.

Dark Current

The bandgap in a bulk heterojunction (BHJ) is delimited by the energy difference between the $LUMO_A$ and $HOMO_D$. Taking this definition into consideration and the photocurrent generation process in organic semiconductors already described, the operation of a OPD is defined by the working principle of a photodiode under applied reverse bias conditions, where in the absence of light, only a very small injected current from the electrodes can flow (i.e. the so-called dark current), as shown in the figure and under illumination, the photogenerated carriers drift under strong electric fields to the respective electrodes.

This internal electric field that is driven by the reverse bias applied to the photodetector is larger in magnitude than the built-in field. This characteristic yields to quick charges sweeps, decreasing the device response time as well as an enhanced photosensitivity.

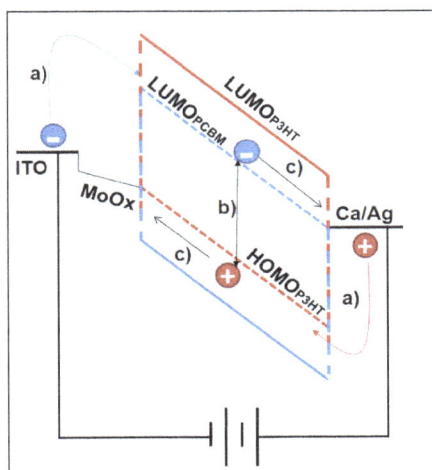

Simplified energy band diagram of an OPD reverse biased. a) Injection of electrons from anode and electrons from cathode, b)carrier recombination, c) collection of holes from the anode and electrons from the cathode.

Ideally, the dark current should be zero. However, it is not negligible due to carrier recombination at the donor/acceptor interface either by: diffusion of electrons or holes across the barrier, or by recombination of electron-hole pairs. Recombination can occur either via charge transfer (CT) state, or via the formation of an exciton by promotion of one of the carriers (the promoter carrier is a recombination particle that has been thermally activated, so it can generate an exciton once it releases the excess of energy. For that reason, the dark current generation is considered a temperature dependent mechanism).

Quantum Efficiency

Quantum efficiency defines the electrical sensitivity to light. It is measured over a range of different wavelengths to characterize the efficiency at each photon energy level. There are two common types: external and internal.

The external quantum efficiency is defined as the number of electron-hole pairs collected that contribute to the photocurrent per incident photon, thus, it is dependent on the wavelength of the incident photon.

$$EQE(\lambda) = \frac{I_{ph}}{P} \cdot \frac{hc}{q\lambda}$$

Where: I_{ph} is the photocurrent, P the incident light power, h the Plancks constant, c the speed of light, λ the wavelength considered and q the elementary charge.

The internal quantum efficiency only considers the photons that are actually absorbed, so, it is described by the ratio of carrier generation rate to the number of photons absorbed by the semiconductor material. In an ideal case, the number of absorbed photons is the same as the number of incoming photons, so, the following relation can be expressed:

$$IQE(\lambda) = \frac{EQE(\lambda)}{A(\lambda)} = \frac{EQE(\lambda)}{1 - r(\lambda) - T(\lambda)}$$

Being A(λ), r(λ), and T(λ) the absorbance, reflectance and transmittance respectively.

Signal to Noise Ratio

The signal-to-noise ratio measures the relation between an arbitrary signal level (not necessarily the most powerful signal possible) and noise. In other terms, signal-to-noise ratio compares the level of a signal to the level of background noise. The higher the ratio, the less interfering the background noise is. It can be expressed by the equation below:

$$SNR = \frac{P_{signal}}{P_{noise}}$$

Here, the signal-to-noise ratio is considered as the relation between the photocurrent (signal under illumination), labeled as Ion and the dark current (signal in the absence of illumination), named as *Ioff* at a reversed bias of -2 V (reference bias as for the applied to photodetectors in imaging applications).

ORGANIC SEMICONDUCTOR

Organic semiconductors are solids whose building blocks are pi-bonded molecules or polymers made up by carbon and hydrogen atoms and – at times – heteroatoms such as nitrogen, sulfur and oxygen. They exist in form of molecular crystals or amorphous thin films. In general, they are electrical insulators, but become semiconducting when charges are either injected from appropriate electrodes, upon doping or by photoexcitation.

General Properties

In molecular crystals the energetic separation between the top of the valence band and the bottom conduction band, i.e. the band gap, is typically 2.5–4 eV, while in inorganic semiconductors the band gaps are typically 1–2 eV. This implies that they are, in fact, insulators rather than semiconductors in the conventional sense. They become semi-conducting only when charge carriers are either injected from the electrodes or generated by intentional or unintentional doping. Charge carriers can also be generated in the course of optical excitation. It is important to realize, however, that the primary optical excitations are neutral excitons with a Coulomb-binding energy of typically 0.5–1.0 eV. The reason is that in organic semiconductors their dielectric constants are as low as 3–4. This impedes efficient photogeneration of charge carriers in neat systems in the bulk. Efficient photogeneration can only occur in binary systems due to charge transfer between donor and acceptor moieties. Otherwise neutral excitons decay radiatively to the ground state – thereby emitting photoluminescence – or non-radiatively. The optical absorption edge of organic semiconductors is typically 1.7–3 eV, equivalent to a spectral range from 700 to 400 nm (which corresponds to the visible spectrum).

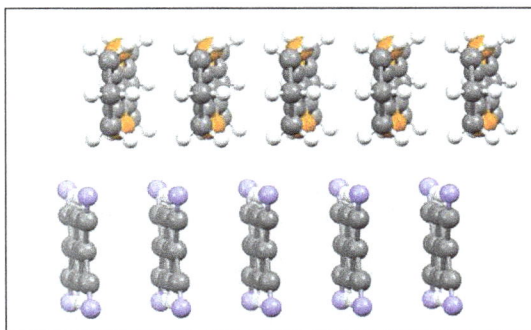

Edge-on view of portion of crystal structure of hexamethyleneTTF/TCNQ charge-transfer salt, highlighting the segregated stacking.

In 1862, Henry Letheby obtained a partly conductive material by anodic oxidation of aniline in sulfuric acid. The material was probably polyaniline. In the 1950s, researchers discovered that polycyclic aromatic compounds formed semi-conducting charge-transfer complex salts with halogens. In particular, high conductivity of 0.12 S/cm was reported in perylene–iodine complex in 1954. This finding indicated that organic compounds could carry current.

The fact that organic semiconductors are, in principle, insulators but become semi-conducting when charge carriers are injected from the electrode(s) was discovered by Kallmann and Pope. They found that a hole current can flow through an anthracene crystal contacted with a positively biased electrolyte containing iodine that can act as a hole injector. This work was stimulated by the earlier discovery by Akamatu et al. That aromatic hydrocarbons become conductive when blended with molecular iodine because a charge-transfer complex is formed. Since it was readily realized that the crucial parameter that controls injection is the work function of the electrode, it was straightforward to replace the electrolyte by a solid metallic or semiconducting contact with an appropriate work function. When both electrons and holes are injected from opposite contacts, they can recombine radiatively and emit light (electroluminescence). It was observed in organic crystals in 1965 by Sano et al.

In 1972, researchers found metallic conductivity in the charge-transfer complex TTF-TCNQ. Superconductivity in charge-transfer complexes was first reported in the Bechgaard salt $(TMTSF)_2PF_6$ in 1980.

In 1973 Dr. John McGinness produced the first device incorporating an organic semiconductor. This occurred roughly eight years before the next such device was created. The "melanin (polyacetylenes) bistable switch" currently is part of the chips collection of the Smithsonian Institution.

An organic polymer voltage-controlled switch from 1974.
Now in the Smithsonian Chip collection.

In 1977, Shirakawa et al. reported high conductivity in oxidized and iodine-doped polyacetylene. They received the 2000 Nobel prize in Chemistry for "The discovery and development of conductive polymers". Similarly, highly conductive polypyrrole was rediscovered in 1979.

Rigid-backbone organic semiconductors are now used as active elements in optoelectronic devices such as organic light-emitting diodes (OLED), organic solar cells, organic field-effect transistors (OFET), electrochemical transistors and recently in biosensing applications. Organic semiconductors have many advantages, such as easy fabrication, mechanical flexibility, and low cost.

The discovery by Kallman and Pope paved the way for applying organic solids as active elements in semiconducting electronic devices, such as organic light-emitting diodes (OLEDs) that rely on the recombination of electrons and hole injected from "ohmic" electrodes, i.e. electrodes with unlimited supply of charge carriers. The next major step towards the technological exploitation of the phenomenon of electron and hole injection into a non-crystalline organic semiconductor was the work by Tang and Van Slyke. They showed that efficient electroluminescence can be generated in a vapor-deposited thin amorphous bilayer of an aromatic diamine (TAPC) and Alq3 sandwiched between an indium-tin-oxide (ITO) anode and an Mg:Ag cathode. Another milestone towards the development of organic light-emitting diodes (OLEDs) was the recognition that also conjugated polymers can be used as active materials. The efficiency of OLEDs was greatly improved when realizing that phosphorescent states (triplet excitons) may be used for emission when doping an organic semiconductor matrix with a phosphorescent dye, such as complexes of iridium with strong spin–orbit coupling.

Work on conductivity of anthracene crystals contacted with an electrolyte showed that optically excited dye molecules adsorbed at the surface of the crystal inject charge carriers. The underlying phenomenon is called sensitized photoconductivity. It occurs when photo-exciting a dye molecule with appropriate oxidation/reduction potential adsorbed at the surface or incorporated in the bulk. This effect revolutionized electrophotography, which is the technological basis of today's office copying machines. It is also the basis of organic solar cells (OSCs), in which the active element is an electron donor, and an electron acceptor material is combined in a bilayer or a bulk heterojunction.

Doping with strong electron donor or acceptors can render organic solids conductive even in the absence of light. Examples are doped polyacetyleneand doped light-emitting diodes. Today organic semiconductors are used as active elements in organic light-emitting diodes (OLEDs), organic solar cells (OSCs) and organic field-effect transistors (OFETs).

Materials

Amorphous Molecular Films

Amorphous molecular films are produced by evaporation or spin-coating. They have been investigated for device applications such as OLEDs, OFETs, and OSCs. Illustrative materials are tris(8-hydroxyquinolinato)aluminium, C_{60}, phenyl-C61-butyric acid methyl ester (PCBM), pentacene, carbazoles, and phthalocyanine.

Molecularly Doped Polymers

Molecularly doped polymers are prepared by spreading a film of an electrically inert polymer, e.g. polycarbonate, doped with typically 30% of charge transporting molecules, on a base electrode. Typical materials are the triphenylenes. They have been investigated for use as photoreceptors in electrophotography. This requires films have a thickness of several micrometers that can be prepared using the doctor-blade technique.

Molecular Crystals

In the early days of fundamental research into organic semiconductors the prototypical materials were free-standing single crystals of the acene family, e.g. anthracene and tetracene. The advantage of employing molecular crystals instead of amorphous film is that their charge carrier mobilities are much larger. This is of particular advantage for OFET applications. Examples are thin films of crystalline rubrene prepared by hot wall epitaxy.

Neat Polymer Films

They are usually processed from solution employing variable deposition techniques including simple spin-coating, ink-jet deposition or industrial reel-to-reel coating which allows preparing thin films on a flexible substrate. The materials of choice are conjugated polymers such as poly-thiophene, poly-phenylenevinylene, and copolymers of alternating donor and acceptor units such as members of the poly(carbazole-dithiophene-benzothiadiazole (PCDTBT) family. For solar cell applications they can be blended with C60 or PCBM as electron acceptors.

Aromatic Short Peptides Self-assemblies

Aromatic short peptides self-assemblies are a kind of promising candidate for bioinspired and durable nanoscale semiconductors. The highly ordered and directional intermolecular π-π interactions and hydrogen-bonding network allow the formation of quantum confined structures within the peptide self-assemblies, thus decreasing the band gaps of the superstructures into semiconductor regions. As a result of the diverse architectures and ease of modification of peptide self-assemblies, their semiconductivity can be readily tuned, doped, and functionalized. Therefore, this family of electroactive supramolecular materials may bridge the gap between the inorganic semiconductor world and biological systems.

Characterization

To design and characterize organic semiconductors used for optoelectronic applications one should first measure the absorption and photoluminescence spectra using commercial instrumentation. However, in order to find out if a material acts as an electron donor or acceptor one has to determine the energy levels for hole and electron transport. The easiest way of doing this, is to employ cyclic voltammetry. However, one has to take into account that using this technique the experimentally determined oxidation and reduction potential are lower bounds because in voltammetry the radical cations and anions are in a polar fluid solution and are, thus, solvated. Such a solvation effect is absent in a solid specimen. The relevant technique to energetically locate the hole transporting states in a solid sample is UV-photoemission spectroscopy. The equivalent technique for electron states is inverse photoemission.

There are several techniques to measure the mobility of charge carriers. The traditional technique is the so-called time of flight (TOF) method. Since this technique requires relatively thick samples it is not applicable to thin films. Alternatively, one can extract the charge carrier mobility from the current flowing in a field effect transistor as a function of both the source-drain and the gate voltage. One should be aware, though, that the FET-mobility is significantly larger than the TOF mobility because of the charge carrier concentration in the transport channel of a FET. Other ways to determine the charge carrier mobility involves measuring space charge limited current (SCLC) flow and "carrier extraction by linearly increasing voltage (CELIV).

In order to characterize the morphology of semiconductor films, one can apply atomic force microscopy (AFM) scanning electron microscopy (SEM) and Grazing-incidence small-angle scattering (GISAS).

Charge Transport

In contrast to organic crystals investigated in the 1960-70s, organic semiconductors that are nowadays used as active media in optoelectronic devices are usually more or less disordered. Combined with the fact that the structural building blocks are held together by comparatively weak van der Waals forces this precludes charge transport in delocalized valence and conduction bands. Instead, charge carriers are localized at molecular entities, e.g. oligomers or segments of a conjugated polymer chain and move by incoherent hopping among adjacent sites with statistically variable energies. Quite often the site energies feature a Gaussian distribution. Also the hopping distances can vary statistically (positional disorder). A consequence of the energetic broadening of the density of states (DOS) distribution is that charge motion is both temperature and field dependent and the charge carrier mobility can be several orders of magnitude lower than in an equivalent crystalline system. This disorder effect on charge carrier motion is diminished in organic field-effect transistors because current flow is confined in a thin layer. Therefore, the tail states of the DOS distribution are already filled so that the activation energy for charge carrier hopping is diminished. For this reason the charge carrier mobility inferred from FET experiments is always higher than that determined from TOF experiments.

In organic semiconductors charge carriers couple to vibrational modes and are referred to as polarons. Therefore, the activation energy for hopping motion contains an additional term due to structural site relaxation upon charging a molecular entity. It turns out, however, that usually the disorder contribution to the temperature dependence of the mobility dominates over the polaronic contribution.

ORGANIC PHOTODETECTOR FABRICATION

The layered stack of photodetectors having a standard geometry is shown in the figure, in which two additional layers are included in order to aid the injection or transport of carriers to the respective electrodes.

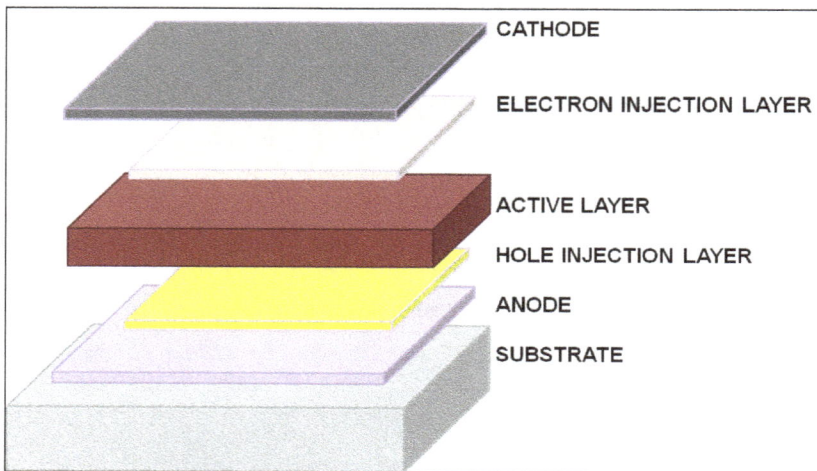

Schematic layout of an organic photodiode.

Materials

Substrate

The most common used substrate types can be divided into: rigid such as glass or silicon and plastics such as Polyphenylene sulfide (PPS), Polyetherimide (PEI), Polyimide (PI) and Polyethylene terephthalate (PET) for flexible applications.

Glass substrates of 3×3 cm^2 with pre-patterned ITO bottom contacts were used.

Organic Semiconductors

The organic semiconductors are the main agent in the device. For this reason, a film formed by these materials is named as the active layer. A solution based active layer of the OPDs was used as a blend of the donor (polymer) and acceptor (fullerene) materials dissolved and mixed in a common solvent (1,2 dichlorobenzene).

The donor is the light absorbing material, in which an electron is excited and transferred to the acceptor. In this project, P3HT was used for this purpose. Poly(3-hexylthiophene), P3HT, is one of the most studied polymer for polymer solar cells as it is easy to synthesize and process (low cost material) as well as reasonably stable, reaching typical efficiency values between 3-5%. P3HT has a bandgap of 2.1 eV. It shows an absorption peak (using 1,2 dichlorobenzene) at $\lambda_{max} = 556\,nm$. Hole mobility is between $10^{-5}\ cm^2 (V\ s)^{-1}$ and $10^{-2}\ cm^2 (V\ s)^{-1}$ with an electron mobility in the range of $6 \times 10^{-4}\ cm^2 (V\ s)^{-1}$ to $1.5 \times 10^{-4}\ cm^2 (V\ s)^{-1}$.

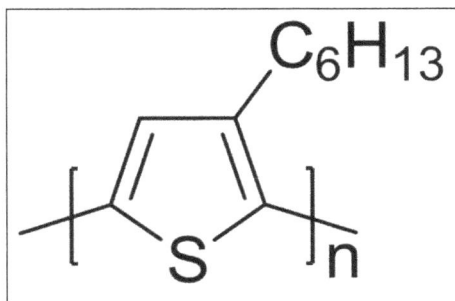

P3HT chemical structure.

A broadly used components as electron acceptors are the derivatives of the buckminsterfullerene (e.g. C60, C61) prepared as Phenyl-C61-butyric acid methyl ester, PCBM, providing good quality blends with donor polymers. Furthermore, the band gap can be tuned from 1.8 eV to 2.2 eV by electron irradiation and the electron mobilities range from $2 \times 10^{-3}\ cm^2 (V\ s)^{-1}$ to $2 \times 10^{-2}\ cm^2 (V\ s)^{-1}$.

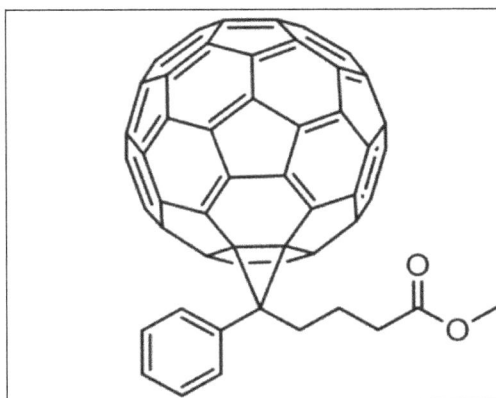

PC60BM chemical structure.

Charge Injection Layers

A selective charge injection layer on each side of the active layer is necessary for allowing only the transit of either electrons (EIL)[1] or holes (HIL)[2]. These layers are based on materials with the ability of transfer either electrons or holes because of the appropriate positioning of their energy levels.

On one side, examples of hole injection materials include the polymer based poly (3,4-ethylenedioxythiophene) polystyrene sulfonate (PEDOT:PSS) and a transition metal oxide such as molybdeenum oxide (MoO_3) with a band gap of 3 - 3.3 eV). On the other side, examples of often used electron injection layers are metal oxides such as zinc oxide (ZnO) and titanium oxide (TiO).

Because of the geometry of the stack of the OPDs fabricated here, in addition to the proper alignment of the energy levels, the selected HIL was Molybdeenum Oxide (MoO_3). Moreover, a EIL was not needed because the LUMOP CBM is at the same energy level of the top metal contact.

Metal Contacts

The metal-organic interface is crucial to enhance the carrier extraction. Moreover, in a standard geometry, the cathode electrode is required to have a low work function in order to generate a built-in potential with the high work function anode electrode.

The variety of choices are metals such as calcium (Ca), silver (Ag) and aluminum (Al) and oxides such as Indium Tin Oxide (ITO), taking into consideration that at least one of the electrodes being transparent such that the light can pass through and reach the active layer. The most commonly used anode electrode material is indium tin oxide (ITO), due to its high optical transmission (>85% on glass), low resistance (<10 Ohm/sq) and work function of $\Phi_{ITO} = 4.7\ eV$.

ITO was used as anode in this work. The cathode was Ca/Ag ($\Phi_{Ca} = 2.9 e$ V and $\Phi_{Ag} = 4.3 e$ V). Silver is used for protection of the low work function Calcium, to prevent oxidation and enhance conductivity.

Finally, a cross-section of a structure schematic and an energy flat band diagram of the fabricated photodetectors are illustrated in the figure.

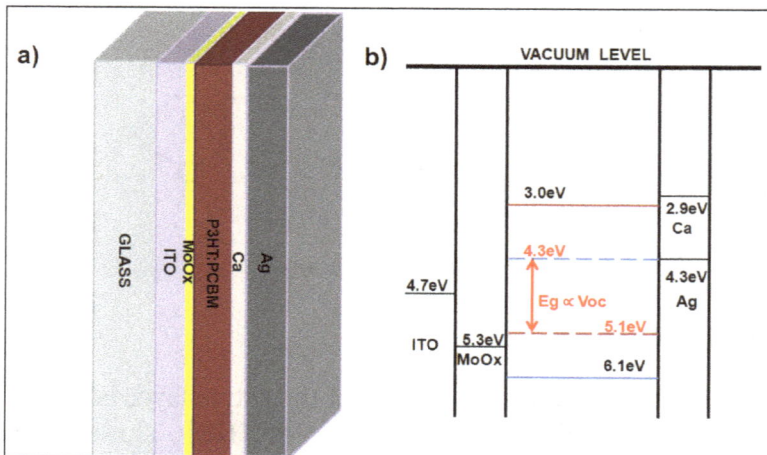

Cross-section structure and b) Energy flat band diagram of the
OPDs based on P3HT:PCBM blends.

OPD Fabrication Process

The fabricated photodetectors had three different structures: reference, bottom patterned and fully patterned.

Reference Structure

The devices with this structure have a layered stack. In this case, the device active area is 0.13 cm², delimited by the superposition of the top and bottom contacts. It should be noted that there are twelve devices in one single sample.

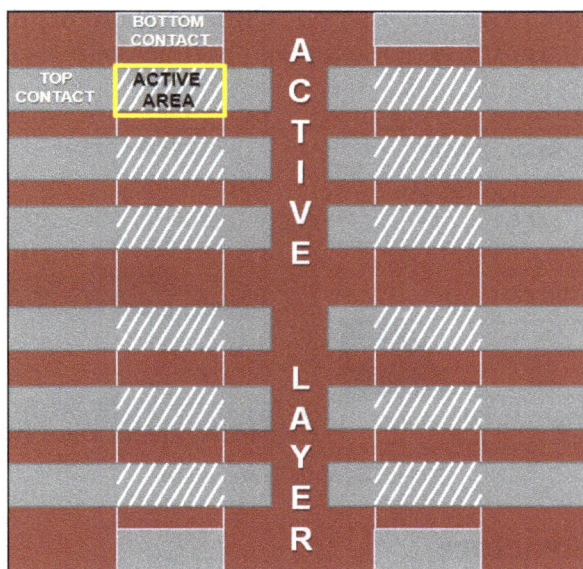

Top view of a sample with twelve devices having a reference structure, with the indication of the active area.

The fabrication process starts with cleaning of commercial 3×3 cm² glass substrate with pre-patterned 140 nm thick ITO in ultrasonic baths with soap, water, acetone and finally isopropanol. Subsequently, 10 nm of the hole injection layer (MoO_3) is thermally evaporated from a quartz crucible and the pressure is kept under 8x10E-7 Torr during the deposition, using a rate of 1 Å per second.

The active layer is a 1:1 donor-acceptor blend of P3HT and PC60BM with 20 mg/ml or 25 mg/ml concentration in order to select the one that offers the optimum device performance. In all the cases the solvent used was orthodichlorobenzene (ODCB). The blend is spin coated in a nitrogen glovebox at 600 rpm. In order to achieve a favorable morphology, the P3HT:PC60BM layer is slow-dried during 10 minutes under a saturated atmosphere, followed by an annealing step on a hot plate at 130 °C during 10 minutes in order to remove an initially depleted PCBM regions and drive the self-organization and crystallization of PCBM. The thicknesses achieved with these processes are approximately 150 nm and 220 nm for the concentrations of 20 mg/ml and 25 mg/ml, respectively.

Finally, the top contact Ca(20 nm)/Ag(120 nm) is thermally evaporated through a shadow mask from a quartz crucible and the pressure is kept under 8x10E-7 Torr during the deposition, using a rate of 1 Å and 6 Å per second respectively.

Bottom Patterned Structure

In this case, an additional layer is patterned on top of ITO by photolithography: A transparent or opaque edge cover layer (ECL). This insulator material allows changing the size and geometry of the device's active area (3.6). Different geometries were used here: circles, squares and ovals, with overall areas from 0.08 cm² down to 1.95E − 5 cm². In such structures, the contact area of the ITO is limited, therefore, the device active area is delimited by the superposition of the top and bottom contacts across the edge cover layer opening.

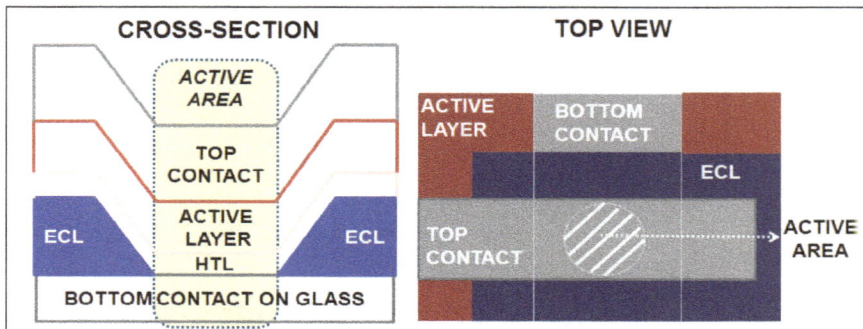

No-Isolated OPD structure.

Fully Patterned Structure

In this structure, the active area is still considered as the stack across the opening delimited by the ECL aperture, but the difference with the previous structure is that the active layer surrounds the ECL opening, so each device is isolated from its neighbors.

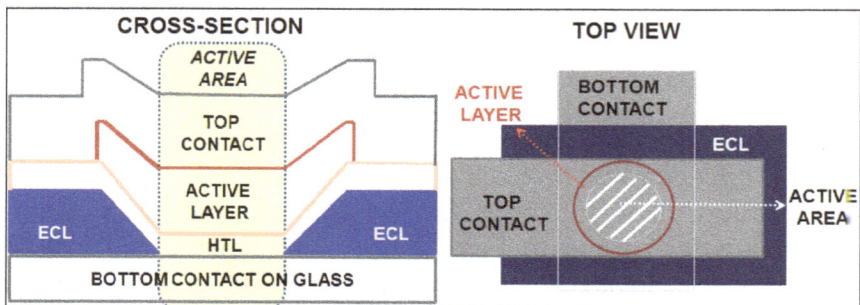

Isolated OPD structure.

The OPDs studied and fabricated here are meant to be used for imaging applications, so high resolution, high aspect ratios and further miniaturization of the photodetectors are required features for this devices. Therefore, patterning is selected as technique to fabricate devices. Patterning can be done with shadow masking, inkjet printing or photolithography. However, the minimum feature size obtained with shadow masking or inkjet printing is limited to 15 - 35 µm whereas feature size smaller than 10 µm can be achieved with lithography. Consequently, patterning with photolithography is used here. Photolithography, forms a resist pattern on the sample through development and transfers the resist pattern onto the substrate as a protective material for etching. The lithography of organic materials presents some issues related to the solvent, photoresist and developer as they are hardly incompatible with the active material.

The process for patterning solution processed layers was developed where first the organic semiconductor layer is coated, then, the photoresist is spin coated. The photoresist is exposed to broadband UV light through a shadow mask to transfer the patterns on the sample. Later, the sample is developed. Next, the sample is exposed to oxygen plasma. After complete removal of the photoresist at locations exposed, the organic semiconductor film is etched away. After the organic layer is removed, it is expected that the remaining mask consists of a thin layer of photoresist, which is then stripped. What remains are the islands of the organic semiconductor.

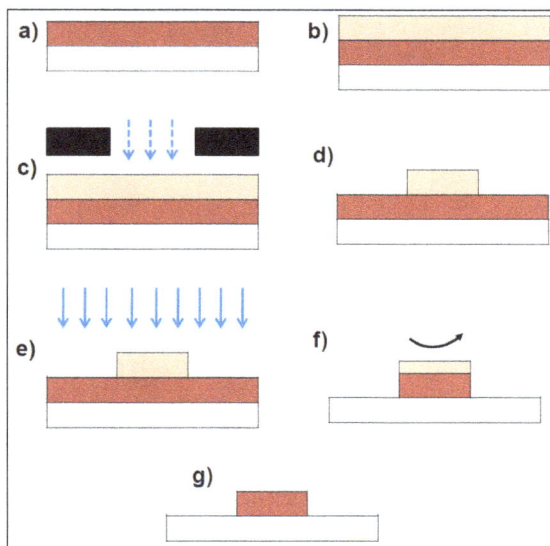

Schematic of the patterning process implemented with photolithography.

This additional patterning step is designed to determine the OPD area that should remain after the device is exposed to deep reactive ion etching. Dry etching with Oxygen plasma at 50W RF power and 20 sccm oxygen flow is used. The etching process is optimized in order to remove the exposed organic film, but at the same time a thin layer of photoresist still remains, protecting the OPD area. The remaining photoresist is stripped and the top contact is evaporated.

An example of the isolated islands of the organic film after etching is illustrated in figure. The inner circle corresponds to the ECL aperture and the outer circle is the active layer. Different designs of the organic semiconductor can be used from large to small islands as compared to the ECL area. This can enable comparison of the contribution of dark current coming from polymer regions closer to top contact edge.

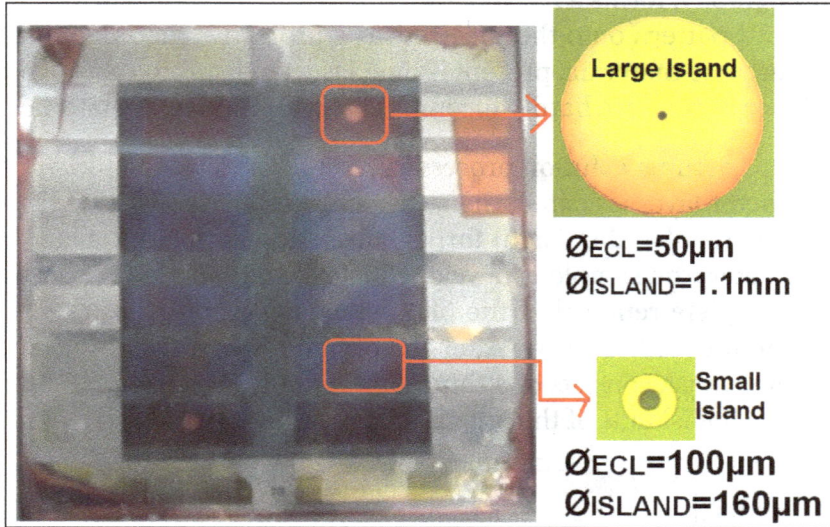

Sample of full patterned photodetectors.

Characterization Methods

Electrical

The current-voltage characteristics were measured under the solar simulator Abet AM1.5G, a one-sun output solar simulator. Light shines upwards into a nitrogen glovebox through a quartz plate. The extracted data is the current in amperes at each applied voltage with and without illumination.

For more accurate measurements in dark, a Cascade Microtech probe station system was used. Finally, the temperature dependence of the electrical characteristics was measured over a temperature range of 300 K to 380 K by adding a thermal chuck to the probe station system.

Optical

To carry out the measurements, a light from xenon and quartz halogen lamps was coupled into a monochromator and their intensities were calibrated with a silicon photodiode. The incident light on the device was chopped and the modulated current signal was detected with a current-voltage and lock-in amplifier.

Furthermore, an optical microscope was used to observe the islands after patterning and etching, and the edge cover layer openings after developing.

Morphological

The topography of organic films that were and were not exposed to the patterning process was studied by atomic force microscopy (AFM) using an Agilent 5100 scanning probe operated in tapping mode. The modular program Gwyddion was used for visualization and analysis of data from the AFM measurements. Furthermore, the cross section of the as-prepared photodetectors with isolated structure were examined by JEOL JSM-6700 field-emission scanning electron microscope. Moreover, the active layer thicknesses were measured using a Dektak 8 Stylus Profiler.

9

Semiconductor Photodetectors

Semiconductor photodetectors consist of active-pixel sensor, cadmium zinc telluride, charge-coupled device, mercury cadmium telluride, light-emitting diode, quantum dot and silicon drift detector. This chapter closely examines about semiconductor photodetectors to provide an extensive understanding of the subject.

A semiconductor material is a continuous crystalline medium characterized by an energy band structure corresponding, in the case of an infinite crystal, to a continuum of states (which, in practice, means that the characteristic dimensions of the crystal are significantly larger than the lattice parameter of the crystal structure; this applies as long as the crystal dimensions are typically larger than a few dozen nanometers). In general terms, the energy structure of a semiconductor consists of a valence band corresponding to molecular bonding states and a conduction band representing the molecular antibonding states.

The energy range lying between the top of the valence band and the bottom of the conduction band is known as the forbidden band, or more commonly the bandgap. An electron situated in the valence band is in a ground state and remains localized to a particular atom in the crystal structure, whereas an electron situated in the conduction band exists in an excited state, in a regime where it interacts very weakly with the crystalline structure. What differentiates semiconductors from insulators is essentially the size of the bandgap: we refer to semiconductors where the bandgap of the material is typically less than or equal to 6 eV, and to insulators when the bandgap is more than 6 eV: above this, the solar spectrum arriving on the Earth's surface is unable to produce inter-band transitions of electrons situated in the valence band of the material. Semiconductor materials are mostly divided into two large classes: elemental semiconductors (group IV of the periodic table): silicon, germanium, diamond, etc. and compound semiconductors: IV-IV (SiC), III-V (GaAs, InP, InSb, GaN) and II-VI (CdTe, ZnSe, ZnS, etc.). Impurities can be introduced into the volume of the semiconductor material and can modify its electrical conduction properties, sometimes considerably. An impurity is known as a donor when it easily releases a free electron into the conduction band. The characteristic energy level of the impurity is therefore in the bandgap, slightly below the conduction band. For example, in the case of compound semiconductors in group IV of the periodic table such as silicon, the main donor impurities are those which, being from group V of the periodic table (arsenic, phosphorous, etc.), are substituted in place of a silicon atom in the crystal structure: since silicon is tetravalent, these atoms naturally form four covalent bonds with the silicon atoms around them, and also easily give up their surplus electron to the crystal structure.

These electrons become free to move, subject to a weak activation energy provided by thermal agitation. In this case we refer to n-type doping. In the case of silicon, a group III element incorporated into the crystal structure of silicon naturally forms three covalent bonds around it, and then completes its own outer-shell electronic structure by capturing an electron from its fourth nearestneighbor silicon atom, again subject to a weak thermal activation energy. Such an impurity is known as an acceptor, and doping with acceptors is known as p-type doping. A hole carrying a positive elementary charge and corresponding to a vacant energy state in the valence band is therefore left in the crystal structure of the silicon. In the case of III-V composites, the donors are mostly atoms from group IV (silicon) substituted in place of group III elements, or group VI elements (S, Se, Te) substituted in place of group V elements, and acceptors are group II (zinc, magnesium) substituted in place of group III elements. In the case of II-VI composites, the most commonlyencountered donors belong to group VII (chlorine, etc.) substituted in place of group VI elements, and acceptors belong to either group I (lithium, etc.) or to group V (nitrogen, arsenic, phosphorous, etc). In this last case, the group V element is substituted in place of a group VI element in the semiconductor crystal structure, whereas group I acceptors are substituted in place of group II elements. The chemical potential, or Fermi energy, of an intrinsic semiconductor (i.e. one free from n and p impurities) is found in the middle of the bandgap of the material. When a moderate n-type doping is added, the Fermi level rises from the middle of the bandgap towards the conduction band, by an increasing amount as the level of doping rises. When the level of n-type doping becomes large, the Fermi level can cross the bottom of the conduction band and be found inside this band (Mott transition). The semiconductor then behaves like a metal and for this reason is called a semi-metal. In this case it is referred to as degenerate. In the case of p-type doping, the semiconductor is said to be degenerate when the Fermi level is below the top of the valence band.

Photodetection with Semiconductors: Basic Phenomena

Photodetection in semiconductors works on the general principle of the creation of electron-hole pairs under the action of light. When a semiconductor material is illuminated by photons of an energy greater than or equal to its bandgap, the absorbed photons promote electrons from the valence band into excited states in the conduction band, where they behave like free electrons able to travel long distances across the crystal structure under the influence of an intrinsic or externally-applied electric field. In addition, the positively-charged holes left in the valence band contribute to electrical conduction by moving from one atomic site to another under the effects of the electric field. In this way the separation of electron-hole pairs generated by the absorption of light gives rise to a photocurrent, which refers by definition to the fraction of the photogenerated free charge-carriers collected at the edges of the material by the electrodes of the photodetecting structure, and whose intensity at a given wavelength is an increasing function of the incident light intensity. On this level we can distinguish between two large categories of photodetectors based on the nature of the electric field, which causes the charge separation of photogenerated electron-hold pairs: photoconductors, which consist of a simple layer of semiconductor simply with two ohmic contacts, where the electric field leading to the collection of the charge-carriers is provided by applying a bias voltage between the contacts at either end, and photovoltaic photodetectors, which use the internal electric field of a p-n or Schottky (metalsemiconductor) junction to achieve the charge separation. This last term covers p-n junction photodetectors (photovoltaic structures consisting of a simple p-n junction, and p-i-n photodetectors which include a thin layer of semiconductor material between the p and n region which is not deliberately doped), as well as all Schottky junction photodetectors (Schottky barrier photodiodes and metalsemiconductor-metal (MSM) photodiodes).

Semiconductor Devices

Photoconductors represent the simplest conceivable type of photodetector: they consist of a finite-length semiconductor layer with an ohmic contact at each end. A fixed voltage of magnitude V_B is applied between the two end contacts, in such a way that a bias current I_B flows through the semiconductor layer, simply following Ohm's law. The active optical surface is formed from the region between the two collection electrodes. When it is illuminated, the photogenerated changes produced under the effect of the applied electric field lead to a photocurrent IPH which is added to the bias current, effectively increasing the conductivity of the device.

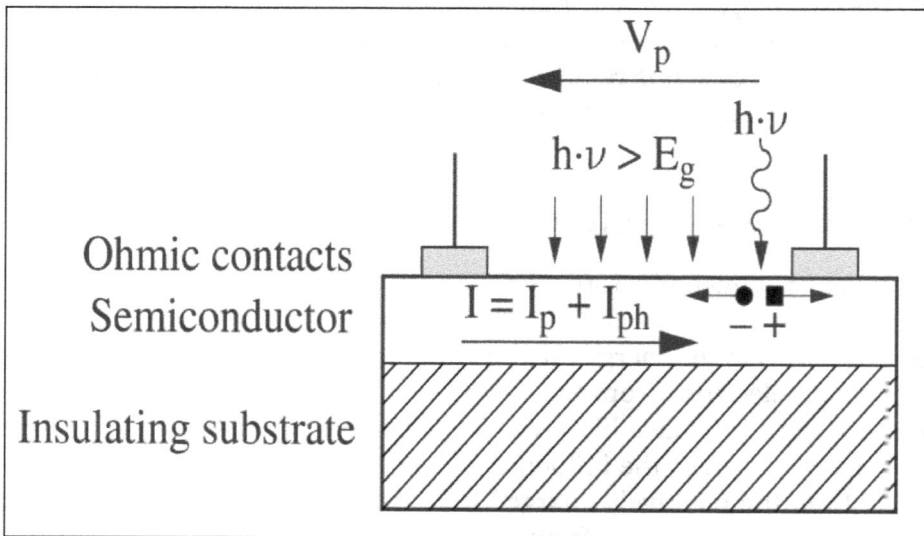

Diagram of a photoconducting device.

The main point of interest in a photoconducting device is its increased gain, the response of photoconductors being typically several orders of magnitude greater than that of photovoltaic detectors for a given material. On the other hand, its other operational parameters (bandwidth, UV/visible contrast, infrared sensitivity) are generally below that of other types of photodetectors, which often greatly limits the scope of its potential applications.

p-n Junctions and P-i-N Structures

In p-n diodes, the metallurgical linkage of a region of a p-type doped semiconductor and a region of n-type doping forms a p-n junction, where the joining of the Fermi levels in equilibrium mostly occurs through a flow of charge between the n and p regions. In equilibrium we therefore find a region with no free charge carriers immediately around the junction, similar to a charged capacitor, where there are, on the n side, positively ionized donors and, on the p side, negatively ionized acceptors (this zone is known as the space charge region (SCR), where ionized donors and acceptors provide fixed charges). The presence of charged donors and acceptors produces an electric field in that region which curves the energy bands and, in equilibrium, forms an energy barrier between the two regions: the bottom of the conduction band and the top of the valence band on the n side are below the corresponding levels on the p side.

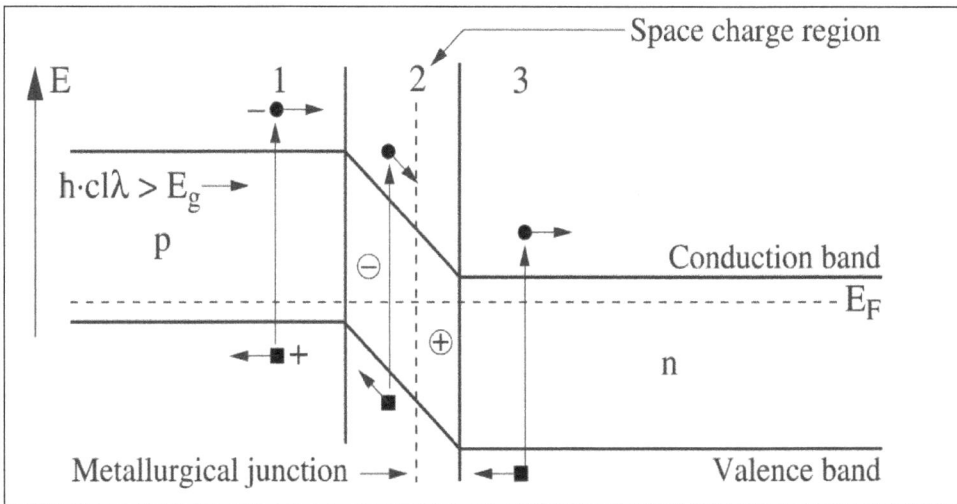

Curvature of the energy bands and mechanisms of photocurrent generation in a p-n junction.

The width of the SCR is a decreasing function of the level of doping in the material, while the height of the energy barrier is an increasing function of it. An electron-hole pair produced in this SCR is therefore separated by the effect of the internal electric field of the junction, and so does not recombine. These are the charge carriers which contribute to the photocurrent, to which we can add, to some extent, those generated at a distance from the junction less than or equal to the diffusion length. The band structure of the junction implies that the photocurrent will consist of minority charge carriers. For this reason, the photocurrent flows in the opposite direction to the bias on the diode, where the forward direction is defined as the direction of flow of the majority charge carriers (from the n to the p region in the case of electrons, and vice versa for holes). Moreover, the application of an opposing external electric field ($V_p - V_n < 0$) allows us to increase the height of the energy barrier in the vicinity of the junction, and also increase the spatial extent of the SCR, which significantly improves the efficiency of the separation of electron-hole pairs by increasing the electric field within the junction.

We note that when the doping level is moderate, the width of the SCR is important. This effect is beneficial in the case of p-n junction photodetectors, where in order to increase the photoresponse it is desirable to ensure that the mechanisms of electron-hole pair generation through incident light take place predominately inside the SCR. A simple means of increasing the spatial extent of the SCR is to introduce between the n and p regions a thin layer of intrinsic semiconductor material which is not intentionally doped: the structure is therefore referred to as p-i-n. Such a structure is interesting because it is possible to maintain high levels of doping in the n and p regions without significantly reducing the extent of the SCR, whose width is then largely determined by the thickness of the "i" layer. Additionally, increasing the width of the SCR reduces the capacitance of the structure, which makes p-i-n structures particularly well-suited for high-speed operation.

Avalanche Effect in P-i-N Structures

When the reverse-bias voltage established at the terminals of a p-in structure increases sufficiently that the electric field established in the junction reaches values close to the breakdown field (in structures of micron-scale thickness, this is generally the case when the bias voltage at the terminals reaches a few dozen volts), the photogenerated charge carriers in the SCR (which is effectively the region that is not intentionally doped) are accelerated enough to separate other secondary charge carriers from the atoms in the lattice that they impact in the course of their motion: this is the avalanche effect which results in a multiplication of the charge carriers in the SCR. The gain is therefore greater than 1 for the generation of charge carriers by light, and this gain can even typically reach 10 or 20 under favorable conditions. This effect is exploited in what are called avalanche photodiodes where the levels of n- and p-type doping are generally adjusted to high values above 10^{18} cm^{-3} to maximize the intrinsic electric field of the junction.

Schottky Junction

A Schottky junction is formed by bringing a metal and a semiconductor into contact. The basic phenomena which lead to the formation of a Schottky junction with an n-type semiconductor are summarized in the figure.

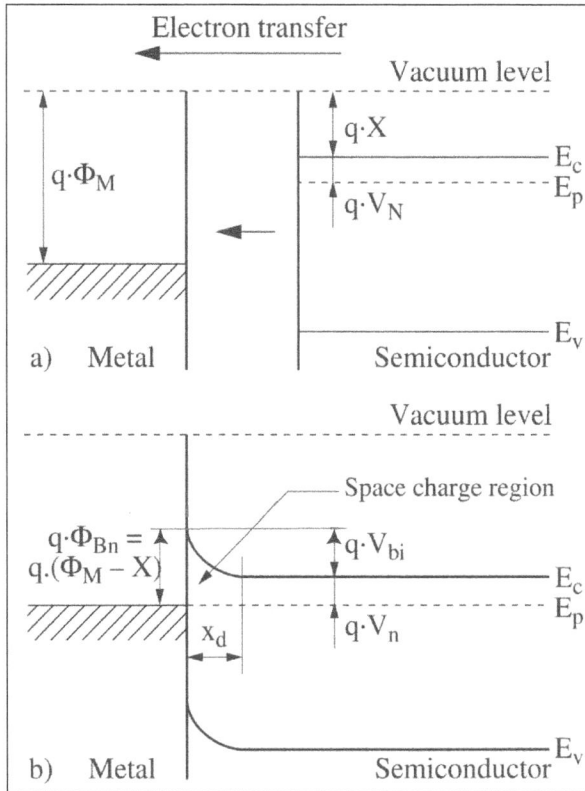

Formation of a Schottky junction (in an n-type semiconductor).

In thermal equilibrium, when the Fermi levels of the metal and the semiconductor are equalized, a transfer of electronic charge occurs from the semiconductor to the metal in the case where the work function $q \cdot \Phi_M$ of the metal (q being the elementary charge) is greater than the electron affinity X of the semiconductor, and a SCR appears at the edge of the semiconductor of width x_d next to the junction, where the only charges present are the positively-ionized donors. A curvature of the energy bands therefore occurs at the junction, which leads to the appearance of an energy barrier between the metal and the semiconductor, called a Schottky barrier, whose height is given to first approximation by the expression:

$$q \cdot \Phi_{Bn} = q \cdot \left(\Phi_M - \chi \right)$$

In equilibrium, therefore, we find an intrinsic electric field immediately next to the metal-semiconductor junction which is comparable in form to that found in a p-n junction. Consequently, it is the phenomenon of photogeneration of charge carriers inside and near to the SCR which is responsible for the appearance of a photocurrent, with the electron-hole pairs being separated by the effect of the electric field in the Schottky junction. It is possible, as in the case of the p-n junction, to modify the intensity of the internal electric field in the junction by applying a bias voltage V between the semiconductor and the metal of the Schottky contact.

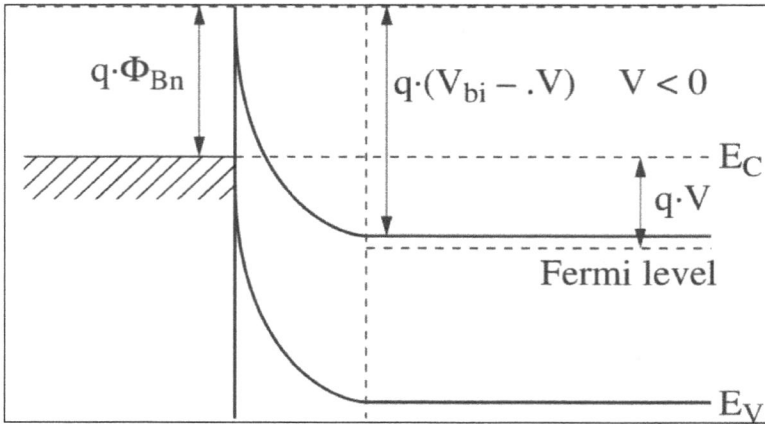

Reverse-bias of a Schottky junction (n-type semiconductor material).

In the case of an n-type semiconductor, the application of a negative voltage between the semiconductor and the metal electrode of the Schottky contact has the effect of reverse-biasing the Schottky junction, which leads to an increase in the height of the effective barrier, along with an increase in the width of the SCR. This last effect is of course favorable for photodetection. Indeed, it follows that the majority charge carriers (electrons) cannot flow towards the Schottky contact, and only the minority carriers (holes) generated by external excitation (in particular photogeneration) can reach the Schottky contact and hence produce an electric current: as in the case of the p-n junction, we therefore find that the current flows in reverse through the Schottky junction, that is, from the semiconductor towards the Schottky contact. The illumination of Schottky photodiodes can occur through the front or rear face (often this second option is chosen in the case where the substrate material is transparent to the light to be detected, as is the case for example with sapphire). In the case of illumination through the front face, we resort to a semi-transparent Schottky contact, characterized by a very small thickness of metal (of the order of 100 Å) selected to ensure sufficient optical transmission: while a thin layer of gold of 100 Å thickness transmits up to 95% of the incident light in the infrared, the percentage transmitted in the ultraviolet is around 30% in the range 300-370 nm. The gain of p-i-n photodiodes (other than the specific case of avalanche photodiodes) and Schottky photodiodes is at most 1, which would be the case if all the photogenerated charge carriers were collected by the electrodes at the ends of the device.

Metal-Semiconductor-Metal (MSM) Structures

An MSM structure consists of two Schottky electrodes, often interlinked in the form of a comb structure, leaving a free semiconductor surface between the two contacts which forms the active region in which light will be absorbed. A bias voltage can be applied between the two electrodes, in order to break the initial electrical symmetry of the contacts: one of the Schottky junctions is reverse-biased, producing a SCR of increased width, and the other junction is forward-biased.

The absorption of light near the reverse-biased junction creates electron-hole pairs which are separated under the effects of the electric field present in the SCR, thus creating the photocurrent. The other electrode, consisting of a forward-biased (and hence transmissive) Schottky junction, simply acts as a collection electrode. The band diagram of the device under increased bias voltage (V_B) is represented schematically in figure, in which L is the distance between two adjacent contact fingers, Φ_0 is the height of the Schottky barrier and I_{ph} is the photocurrent. MSM photodetectors normally use semiconductor materials which are not intentionally doped, are chemically very pure and electrically very resistive. The SCRs associated with Schottky junctions made of these materials are hence of significant width which, for a given bias voltage, allows the electric field of the junction to extend more easily into semiconductor regions some way from the contact. It follows that photogenerated electronhole pairs are more easily separated and collected by the electrodes at either end.

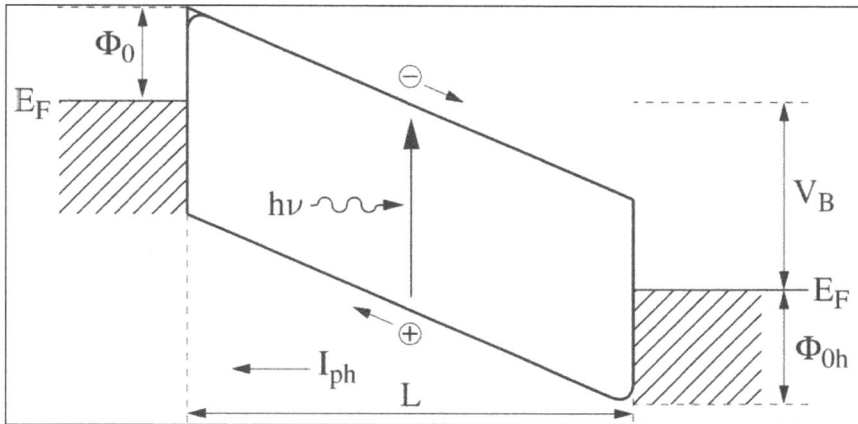

Energy band diagram for an MSM structure under electrical bias; effect of illumination.

Operational Parameters of Photodetectors

The main parameters which define the behavior of an ultraviolet photodetector are respectively the response coefficient, the gain, the quantum efficiency, the bandwidth, the noise equivalent power (NEP) and the detectivity.

Response Coefficient, Gain and Quantum Efficiency

The response coefficient of a photodetector, R_i, links the photocurrent I_{ph} to the power of the incident light P_{opt} through the relationship:

$$I_{ph} = R_i \cdot P_{opt}$$

It is important to note in passing that the response coefficient is a quantity independent of the active optical surface of the photodetector structure: indeed, the photocurrent as well as the incident optical power are both, in the ideal case, proportional to the active optical surface. At a given wavelength λ, the flux Φ of photons arriving on the semiconductor surface, which is defined as the number of photons reaching the active surface per unit time, is given by:

$$\Phi = P_{opt} \cdot \lambda / (h.c)$$

where h is the Planck constant and c is the speed of light.

The quantum efficiency η is defined as the probability of creating an electron-hole pair from an absorbed photon. Considering that all the incident light is absorbed in the semiconductor material, the rate G of electron-hole pair generation per unit time is thus given by:

$$G = \eta \cdot \Phi = \eta \cdot P_{opt} \cdot \lambda / (h \cdot c)$$

If we now introduce the gain parameter g which corresponds to the number of charge carriers detected relative to the number of photogenerated electron-hole pairs, then the photocurrent is given by the equation:

$$I_{ph} = q \cdot G \cdot g = q \cdot \eta \cdot P_{opt} \lambda / (h \cdot c) \cdot g = (q \cdot \eta \cdot \lambda / (hc) \cdot g) \cdot P_{opt}$$

where q is the elementary charge (1.602 x 10⁻¹⁹ C), from which we obtain the expression for the response coefficient of the detector:

$$R_i = q \cdot g \cdot \eta \cdot \lambda / (hc)$$

Temporal Response and Bandwidth

The speed of response of a photodetector may be limited by capacitative effects, by the trapping of charge carriers or by the saturation speed of charge carriers in the semiconductor. These phenomena all lead to a reduction in the response of the photodetector in the high-frequency domain. The cutoff frequency f_C of the photodetector is defined as the frequency of optical signal for which the response coefficient is half that for a continuous optical signal. The temporal response of a photodetector is characterized by the fall time τ_f (or the rise time τ_r), which is defined as the time needed for the photocurrent to fall from 90% to 10% of its maximum (or to rise from 10% to 90% of it). In the case of a transient exponential response with a time constant τ, the following relationship links the bandwidth BW and the temporal response of the photodetector:

$$BP = 1/(2 \cdot \pi \cdot \tau) = 2.2/(2 \cdot \pi \tau_m) = 2.2/(2 \cdot \pi \cdot \tau_d)$$

Noise Equivalent Power

The NEP is defined as the incident optical power for which the signal-to-noise ratio is 1, and hence the photocurrent I_{ph} is equal to the noise current I_b. In other words, it is the smallest optical power which can be measured. It follows that the NEP parameter is given by the equation:

$$NEP = I_b / R_i \ (in\,W)$$

In the case of white noise, the noise current I_b increases as the square root of the bandwidth of the photodetector device. It follows that it is preferable and customary to use the following expression for the NEP, normalized with respect to the bandwidth BW:

$$NEP^* = NEP \cdot (BW)^{-1/2} \left(in\,W \cdot Hz^{-1/2}\right)$$

In semiconductors, there are five sources of noise:

- Shot noise, mainly due to the random nature of the collisions of incident photons;

- Thermal noise, due to random collisions of charge carriers with the atoms of the crystal lattice, in permanent vibration due to thermal motion;

- Partition noise, caused by the separation of the electric current into two parts flowing across separate electrical contacts;

- Generation-recombination noise, caused by the random generation and recombination of charge carriers, either band to band or via trapping levels situated in the bandgap;

- $1/f$ noise, associated with the presence of potential barriers at the level of the electrical contacts. This last type of noise dominates at low frequencies.

Detectivity

This figure of merit is defined by the equation:

$$D = (NEP)^{-1} = R_i / I_b \left(in\,W^{-1}\right)$$

In general terms, the photocurrent signal increases in proportion to the active optical area A_{opt}, and in addition the noise current increases with the square root of the product of the active optical area with the bandwidth BW. It follows that the preferred method of comparing between different photodetectors is to use an expression for the detectivity normalized with respect to these parameters, written:

$$D^* = D \cdot \left(A_{opt} \cdot BP\right)^{1/2} = \left(R_i / I_b\right) \cdot \left(A_{opt} \cdot BP\right)^{1/2} \left(in\,W^{-1} \cdot cm.Hz^{1/2}\right)$$

The normalized detectivity is the most important parameter for characterizing a photodetector because it allows direct comparison of the performance of photodetectors using technologies and methods of operation which are at first glance very different. It is clear from the preceding definitions that the determination of the NEP and the detectivity requires measurement of three parameters: the response coefficient, the bandwidth and the noise current of the photodetector device. The measurement of the noise current must be made in darkness. The device is biased using a very stable voltage source, and the entire measurement system must itself have an intrinsic noise level considerably lower than the intrinsic noise of the photodetector device.

ACTIVE-PIXEL SENSOR

An active-pixel sensor (APS) is an image sensor where each pixel sensor unit cell has a photodetector (typically a pinned photodiode) and one or more active transistors. In a metal–oxide–semiconductor (MOS) active-pixel sensor, MOS field-effect transistors (MOSFETs) are used as amplifiers. There are different types of APS, including the early NMOS APS and the much more common complementary MOS (CMOS) APS, also known as the CMOS sensor, which is widely used in digital camera technologies such as cell phone cameras, web cameras, most modern digital pocket cameras, most digital single-lens reflex cameras (DSLRs), and mirrorless interchangeable-lens cameras (MILCs). CMOS sensors emerged as an alternative to charge-coupled device (CCD) image sensors and eventually outsold them by the mid-2000s.

CMOS image sensor.

The term 'active pixel sensor' is also used to refer to the individual pixel sensor itself, as opposed to the image sensor. In this case, the image sensor is sometimes called an active pixel sensor imager, or active-pixel image sensor.

Background

The basis for modern image sensors is metal–oxide–semiconductor (MOS) technology, which originates from the MOSFET (MOS field-effect transistor) invented by Mohamed M. Atalla and Dawon Kahng at Bell Labs in 1959. They demonstrated two MOSFET fabrication processes in 1960, PMOS (p-type MOS) and NMOS (n-type MOS). Both processes were later combined and adapted into the CMOS (complementary MOS) process by Chih-Tang Sah and Frank Wanlass at Fairchild Semiconductor in 1963.

While researching MOS technology, Willard Boyle and George E. Smith realized that an electric charge could be stored on a tiny MOS capacitor, which became the basic building block of the charge-couple device (CCD), which they invented in 1969. An issue with CCD technology was that it required the need for nearly perfect charge transfer, which, according to Eric Fossum, "makes their radiation 'soft,' difficult to use under low light conditions, difficult to manufacture in large array sizes, difficult to integrate with on-chip electronics, difficult to use at low temperatures, difficult to use at high frame rates, and difficult to manufacture in non-silicon materials that extend wavelength response".

At RCA Laboratories, a research team including Paul K. Weimer, W.S. Pike and G. Sadasiv in 1969 proposed a solid-state image sensor with scanning circuits using thin-film transistors (TFTs), with photoconductive film used for the photodetector. A low-resolution "mostly digital" N-channel MOSFET (NMOS) imager with intra-pixel amplification, for an optical mouse application, was demonstrated by Richard F. Lyon in 1981. Another type of image sensor technology that is related to the APS is the hybrid infrared focal plane array (IRFPA), designed to operate at cryogenic temperatures in the infrared spectrum. The devices are two chips that are put together like a sandwich: one chip contains detector elements made in InGaAs or HgCdTe, and the other chip is typically made of silicon and is used to read out the photodetectors. The exact date of origin of these devices is classified, but they were in use by the mid-1980s.

A key element of the modern CMOS sensor is the pinned photodiode (PPD). It was invented by Nobukazu Teranishi, Hiromitsu Shiraki and Yasuo Ishihara at NEC in 1980, and then publicly reported by Teranishi and Ishihara with A. Kohono, E. Oda and K. Arai in 1982, with the addition of an anti-blooming structure. The pinned photodiode is a photodetector structure with low lag, low noise, high quantum efficiency and low dark current. The new photodetector structure invented at NEC was given the name "pinned photodiode" (PPD) by B.C. Burkey at Kodak in 1984. In 1987, the PPD began to be incorporated into most CCD sensors, becoming a fixture in consumer electronic video cameras and then digital still cameras. Since then, the PPD has been used in nearly all CCD sensors and then CMOS sensors.

Passive-pixel Sensor

The precursor to the APS was the passive-pixel sensor (PPS), a type of photodiode array (PDA). A passive-pixel sensor consists of passive pixels which are read out without amplification, with each pixel consisting of a photodiode and a MOSFET switch. In a photodiode array, pixels contain a p-n junction, integrated capacitor, and MOSFETs as selection transistors. A photodiode array was proposed by G. Weckler in 1968, predating the CCD. This was the basis for the PPS, which had image sensor elements with in-pixel selection transistors, proposed by Peter J.W. Noble in 1968, and by Savvas G. Chamberlain in 1969.

Passive-pixel sensors were being investigated as a solid-state alternative to vacuum-tube imaging devices. The MOS passive-pixel sensor used just a simple switch in the pixel to read out the photodiode integrated charge. Pixels were arrayed in a two-dimensional structure, with an access enable wire shared by pixels in the same row, and output wire shared by column. At the end of each column was a transistor. Passive-pixel sensors suffered from many limitations, such as high noise, slow readout, and lack of scalability. Early photodiode arrays were complex and impractical, requiring selection transistors to be fabricated within each pixel, along with on-chip multiplexer circuits. The noise of photodiode arrays was also a limitation to performance, as the photodiode readout bus capacitance resulted in increased noise level. Correlated double sampling (CDS) could also not be used with a photodiode array without external memory. It was not possible to fabricate active pixel sensors with a practical pixel size in the 1970s, due to limited microlithography technology at the time. Because the MOS process was so variable and MOS transistors had characteristics that changed over time (Vth instability), the CCD's charge-domain operation was more manufacturable than MOS passive pixel sensors.

Active-pixel Sensor

The active-pixel sensor consists of active pixels, each containing one or more MOSFET amplifiers which convert the photo-generated charge to a voltage, amplify the signal voltage, and reduce noise. The first MOS active-pixel sensor was the Charge Modulation Device (CMD) invented by Olympus in Japan during the mid-1980s. This was enabled by advances in MOSFET semiconductor device fabrication, with MOSFET scaling reaching smaller micron and then sub-micron levels during the 1980s to early 1990s. The first MOS APS was fabricated by Tsutomu Nakamura's team at Olympus in 1985. The term *active pixel sensor* (APS) was coined by Nakamura while working on the CMD active-pixel sensor at Olympus. The CMD imager had a vertical APS structure, which increases fill-factor (or reduces pixel size) by storing the signal charge under an output NMOS transistor. Other Japanese semiconductor companies soon followed with their own active pixel sensors during the late 1980s to early 1990s. Between 1988 and 1991, Toshiba developed the "double-gate floating surface transistor" sensor, which had a lateral APS structure, with each pixel containing a buried-channel MOS photogate and a PMOS output amplifier. Between 1989 and 1992, Canon developed the base-stored image sensor (BASIS), which used a vertical APS structure similar to the Olympus sensor, but with bipolar transistors rather than MOSFETs.

In the early 1990s, American companies began developing practical MOS active pixel sensors. In 1991, Texas Instruments developed the bulk CMD (BCMD) sensor, which was fabricated at the company's Japanese branch and had a vertical APS structure similar to the Olympus CMD sensor, but was more complex and used PMOS rather than NMOS transistors.

CMOS Sensor

By the late 1980s to early 1990s, the CMOS process was well-established as a well-controlled stable semiconductor manufacturing process and was the baseline process for almost all logic and microprocessors. There was a resurgence in the use of passive-pixel sensors for low-end imaging applications,while active-pixel sensors began being used for low-resolution high-function applications such as retina simulationand high-energy particle detectors. However, CCDs continued to have much lower temporal noise and fixed-pattern noise and were the dominant technology for consumer applications such as camcorders as well as for broadcast cameras, where they were displacing video camera tubes.

In 1993, the first practical APS to be successfully fabricated outside of Japan was developed at NASA's Jet Propulsion Laboratory (JPL), which fabricated a CMOS compatible APS, with its development led by Eric Fossum. It had a lateral APS structure similar to the Toshiba sensor, but was fabricated with CMOS rather than PMOS transistors.It was the first CMOS sensor with intra-pixel charge transfer.

Fossum, who worked at JPL, led the development of an image sensor that used intra-pixel charge transfer along with an in-pixel amplifier to achieve true correlated double sampling (CDS) and low temporal noise operation, and on-chip circuits for fixed-pattern noise reduction. He also published an extensive 1993 article predicting the emergence of APS imagers as the commercial successor of CCDs. He classified two types of APS structures, the lateral APS and the vertical APS. He also gave an overview of the history of APS technology, from the first APS sensors in Japan to the development of the CMOS sensor at JPL.

In 1994, Fossum proposed an improvement to the CMOS sensor: the integration of the pinned photodiode (PPD). A CMOS sensor with PPD technology was first fabricated in 1995 by a joint JPL and Kodak team that included Fossum along with P. P. K. Lee, R. C. Gee, R. M. Guidash and T. H. Lee.Between 1993 and 1995, the Jet Propulsion Laboratory developed a number of prototype devices, which validated the key features of the technology. Though primitive, these devices demonstrated good image performance with high readout speed and low power consumption.

In 1995, being frustrated by the slow pace of the technology's adoption, Fossum and his then-wife Dr. Sabrina Kemeny co-founded Photobit Corporation to commercialize the technology. It continued to develop and commercialize APS technology for a number of applications, such as web cams, high speed and motion capture cameras, digital radiography, endoscopy (pill) cameras, digital single-lens reflex cameras (DSLRs) and camera-phones. Many other small image sensor companies also sprang to life shortly thereafter due to the accessibility of the CMOS process and all quickly adopted the active pixel sensor approach.

Photobit's CMOS sensors found their way into webcams manufactured by Logitech and Intel, before Photobit was purchased by Micron Technology in 2001. The early CMOS sensor market was initially led by American manufacturers such as Micron, GoPro, and Omnivision, allowing the United States to briefly recapture a portion of the overall image sensor market from Japan, before the CMOS sensor market eventually came to be dominated by Japan, South Korea and China. The CMOS sensor with PPD technology was further advanced and refined by R. M. Guidash in 1997, K. Yonemoto and H. Sumi in 2000, and I. Inoue in 2003. This led to CMOS sensors achieve imaging performance on par with CCD sensors, and later exceeding CCD sensors.

By 2000, CMOS sensors were used in a variety of applications, including low-cost cameras, PC cameras, fax, multimedia, security, surveillance, and videophones.

The video industry switched to CMOS cameras with the advent of high-definition video (HD video), as the large number of pixels would require significantly higher power consumption with CCD sensors, which would overheat and drain batteries.Sony in 2007 commercialized CMOS sensors with an original column A/D conversion circuit, for fast, low-noise performance, followed in 2009 by the CMOS back-illuminated sensor (BI sensor), with twice the sensitivity of conventional image sensors and going beyond the human eye.

CMOS sensors went on to have a significant cultural impact, leading to the mass proliferation of digital cameras and camera phones, which bolstered the rise of social media and selfie culture, and impacted social and political movements around the world. By 2007, sales of CMOS active-pixel sensors had surpassed CCD sensors, with CMOS sensors accounting for 54% of the global image sensor market at the time. By 2012, CMOS sensors increased their share to 74% of the market. As of 2017, CMOS sensors account for 89% of global image sensor sales. In recent years, the CMOS sensor technology has spread to medium-format photography with Phase One being the first to launch a medium format digital back with a Sony-built CMOS sensor.

In 2012, Sony introduced the stacked CMOS BI sensor. Fossum now performs research on the Quanta Image Sensor (QIS) technology. The QIS is a revolutionary change in the way we collect images in a camera that is being invented at Dartmouth. In the QIS, the goal is to count every photon that strikes the image sensor, and to provide resolution of 1 billion or more specialized photoelements (called jots) per sensor, and to read out jot bit planes hundreds or thousands of times per second resulting in terabits/sec of data.

Comparison to CCDs

APS pixels solve the speed and scalability issues of the passive-pixel sensor. They generally consume less power than CCDs, have less image lag, and require less specialized manufacturing facilities. Unlike CCDs, APS sensors can combine the image sensor function and image processing functions within the same integrated circuit. APS sensors have found markets in many consumer applications, especially camera phones. They have also been used in other fields including digital radiography, military ultra high speed image acquisition, security cameras, and optical mice. Manufacturers include Aptina Imaging (independent spinout from Micron Technology, who purchased

Photobit in 2001), Canon, Samsung, STMicroelectronics, Toshiba, OmniVision Technologies, Sony, and Foveon, among others. CMOS-type APS sensors are typically suited to applications in which packaging, power management, and on-chip processing are important. CMOS type sensors are widely used, from high-end digital photography down to mobile-phone cameras.

Advantages of CMOS Compared with CCD

Blooming in a CCD image.

A big advantage of a CMOS sensor is that it is typically less expensive than a CCD sensor. A CMOS sensor also typically has better control of blooming (that is, of bleeding of photo-charge from an over-exposed pixel into other nearby pixels).

In three-sensor camera systems that use separate sensors to resolve the red, green, and blue components of the image in conjunction with beam splitter prisms, the three CMOS sensors can be identical, whereas most splitter prisms require that one of the CCD sensors has to be a mirror image of the other two to read out the image in a compatible order. Unlike CCD sensors, CMOS sensors have the ability to reverse the addressing of the sensor elements. CMOS Sensors with a film speed of ISO 4 million exist.

Disadvantages of CMOS Compared with CCD

Distortion caused by a rolling shutter.

Since a CMOS sensor typically captures a row at a time within approximately 1/60th or 1/50th of a second (depending on refresh rate) it may result in a "rolling shutter" effect, where the image is skewed (tilted to the left or right, depending on the direction of camera or subject movement). For example, when tracking a car moving at high speed, the car will not be distorted but the background will appear to be tilted. A frame-transfer CCD sensor or "global shutter" CMOS sensor does not have this problem, instead captures the entire image at once into a frame store.

The active circuitry in CMOS pixels takes some area on the surface which is not light-sensitive, reducing the photon-detection efficiency of the device (back-illuminated sensors can mitigate this problem). But the frame-transfer CCD also has about half non-sensitive area for the frame store nodes, so the relative advantages depend on which types of sensors are being compared.

Architecture

Pixel

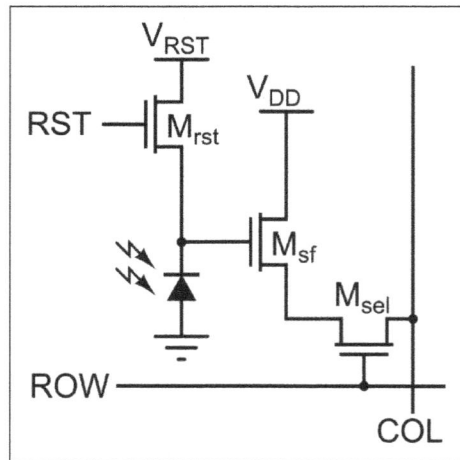

A three-transistor active pixel sensor.

The standard CMOS APS pixel today consists of a photodetector (pinned photodiode), a floating diffusion, and the so-called 4T cell consisting of four CMOS (complementary metal–oxide–semiconductor) transistors, including a transfer gate, reset gate, selection gate and source-follower readout transistor. The pinned photodiode was originally used in interline transfer CCDs due to its low dark current and good blue response, and when coupled with the transfer gate, allows complete charge transfer from the pinned photodiode to the floating diffusion (which is further connected to the gate of the read-out transistor) eliminating lag. The use of intrapixel charge transfer can offer lower noise by enabling the use of correlated double sampling (CDS). The Noble 3T pixel is still sometimes used since the fabrication requirements are less complex. The 3T pixel comprises the same elements as the 4T pixel except the transfer gate and the photodiode. The reset transistor, M_{rst}, acts as a switch to reset the floating diffusion to V_{RST}, which in this case is represented as the gate of the M_{sf} transistor. When the reset transistor is turned on, the photodiode is effectively connected to the power supply, V_{RST}, clearing all integrated charge. Since the reset transistor is n-type, the pixel operates in soft reset. The read-out transistor, M_{sf}, acts as a buffer (specifically, a source follower), an amplifier which allows the pixel voltage to be observed without removing the accumulated charge. Its power supply, V_{DD}, is typically tied to the power supply of the reset transistor V_{RST}. The select transistor, M_{sel}, allows a single row of the pixel array to be read by the read-out electronics. Other innovations of the pixels such as 5T and 6T pixels also exist. By adding extra transistors, functions such as global shutter, as opposed to the more common rolling shutter, are possible. In order to increase the pixel densities, shared-row, four-ways and eight-ways shared read out, and other architectures can be employed. A variant of the 3T active pixel is the Foveon X3 sensor invented by Dick Merrill. In this device, three photodiodes are stacked on top of each other using planar fabrication techniques, each photodiode having its own 3T circuit. Each successive layer acts as a filter for the layer below it shifting the spectrum of absorbed light in successive layers. By deconvolving the response of each layered detector, red, green, and blue signals can be reconstructed.

Array

A typical two-dimensional array of pixels is organized into rows and columns. Pixels in a given row share reset lines, so that a whole row is reset at a time. The row select lines of each pixel in a row are tied together as well. The outputs of each pixel in any given column are tied together. Since only one row is selected at a given time, no competition for the output line occurs. Further amplifier circuitry is typically on a column basis.

Size

The size of the pixel sensor is often given in height and width, but also in the optical format.

Lateral and Vertical Structures

There are two types of active-pixel sensor (APS) structures, the lateral APS and vertical APS. Eric Fossum defines the lateral APS as follows:

> A lateral APS structure is defined as one that has part of the pixel area used for photodetection and signal storage, and the other part is used for the active transistor(s). The advantage of this approach, compared to a vertically integrated APS, is that the fabrication process is simpler, and is highly compatible with state-of-the-art CMOS and CCD device processes.

Fossum defines the vertical APS as follows:

> A vertical APS structure increases fill-factor (or reduces pixel size) by storing the signal charge under the output transistor.

Thin-film Transistors

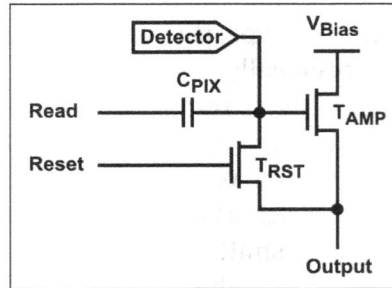

A two-transistor active/passive pixel sensor.

For applications such as large-area digital X-ray imaging, thin-film transistors (TFTs) can also be used in APS architecture. However, because of the larger size and lower transconductance gain of TFTs compared with CMOS transistors, it is necessary to have fewer on-pixel TFTs to maintain image resolution and quality at an acceptable level. A two-transistor APS/PPS architecture has been shown to be promising for APS using amorphous silicon TFTs. In the two-transistor APS architecture on the right, T_{AMP} is used as a switched-amplifier integrating functions of both M_{sf} and M_{sel} in the three-transistor APS. This results in reduced transistor counts per pixel, as well as increased pixel transconductance gain. Here, C_{pix} is the pixel storage capacitance, and it is also used to capacitively couple the addressing pulse of the "Read" to the gate of T_{AMP} for ON-OFF switching. Such pixel readout circuits work best with low capacitance photoconductor detectors such as amorphous selenium.

Design Variants

Many different pixel designs have been proposed and fabricated. The standard pixel is the most common because it uses the fewest wires and the fewest, most tightly packed transistors possible for an active pixel. It is important that the active circuitry in a pixel take up as little space as possible to allow more room for the photodetector. High transistor count hurts fill factor, that is, the percentage of the pixel area that is sensitive to light. Pixel size can be traded for desirable qualities such as noise reduction or reduced image lag. Noise is a measure of the accuracy with which the incident light can be measured. Lag occurs when traces of a previous frame remain in future frames, i.e. the pixel is not fully reset. The voltage noise variance in a soft-reset (gate-voltage regulated) pixel is $V_n^2 = kT/2C$, but image lag and fixed pattern noise may be problematic. In rms electrons, the noise is $N_e = \dfrac{\sqrt{kTC/2}}{q}$.

Hard Reset

Operating the pixel via hard reset results in a Johnson–Nyquist noise on the photodiode of $V_n^2 = kT/C$ or $N_e = \dfrac{\sqrt{kTC}}{q}$, but prevents image lag, sometimes a desirable tradeoff. One way to use hard reset is replace M_{rst} with a p-type transistor and invert the polarity of the RST signal. The presence of the p-type device reduces fill factor, as extra space is required between p- and n-devices; it also removes the possibility of using the reset transistor as an overflow anti-blooming drain, which is a commonly exploited benefit of the n-type reset FET. Another way to achieve hard reset, with the n-type FET, is to lower the voltage of V_{RST} relative to the on-voltage of RST. This reduction may reduce headroom, or full-well charge capacity, but does not affect fill factor, unless V_{DD} is then routed on a separate wire with its original voltage.

Combinations of Hard and Soft Reset

Techniques such as flushed reset, pseudo-flash reset, and hard-to-soft reset combine soft and hard reset. The details of these methods differ, but the basic idea is the same. First, a hard reset is done, eliminating image lag. Next, a soft reset is done, causing a low noise reset without adding any lag. Pseudo-flash reset requires separating V_{RST} from V_{DD}, while the other two techniques add more complicated column circuitry. Specifically, pseudo-flash reset and hard-to-soft reset both add transistors between the pixel power supplies and the actual V_{DD}. The result is lower headroom, without affecting fill factor.

Active Reset

A more radical pixel design is the active-reset pixel. Active reset can result in much lower noise levels. The tradeoff is a complicated reset scheme, as well as either a much larger pixel or extra column-level circuitry.

CADMIUM ZINC TELLURIDE

Cadmium zinc telluride, (CdZnTe) or CZT, is a compound of cadmium, zinc and tellurium or, more strictly speaking, an alloy of cadmium telluride and zinc telluride. A direct bandgap semiconductor, it is used in a variety of applications, including semiconductor radiation detectors, photorefractive gratings, electro-optic modulators, solar cells, and terahertz generation and detection. The band gap varies from approximately 1.4 to 2.2 eV, depending on composition.

Radiation detectors using CZT can operate in direct-conversion (or photoconductive) mode at room temperature, unlike some other materials (particularly germanium) which require liquid nitrogen cooling. Their relative advantages include high sensitivity for x-rays and gamma-rays, due to the high atomic numbers of Cd and Te, and better energy resolution than scintillator detectors. CZT can be formed into different shapes for different radiation-detecting applications, and a variety of electrode geometries, such as coplanar grids and small pixel detectors, have been developed to provide unipolar (electron-only) operation, thereby improving energy resolution.

CHARGE-COUPLED DEVICE

A specially developed CCD in a wire-bonded package used for ultraviolet imaging.

A charge-coupled device (CCD) is a device for the movement of electrical charge, usually from within the device to an area where the charge can be manipulated, such as conversion into a digital value. This is achieved by "shifting" the signals between stages within the device one at a time. CCDs move charge between capacitive *bins* in the device, with the shift allowing for the transfer of charge between bins.

CCD is a major technology for digital imaging. In a CCD image sensor, pixels are represented by p-doped metal–oxide–semiconductor (MOS) capacitors. These MOS capacitors, the basic building blocks of a CCD, are biased above the threshold for inversion when image acquisition begins, allowing the conversion of incoming photons into electron charges at the semiconductor-oxide interface; the CCD is then used to read out these charges. Although CCDs are not the only technology to allow for light detection, CCD image sensors are widely used in professional, medical, and scientific applications where high-quality image data are required. In applications with less exacting quality demands, such as consumer and professional digital cameras, active pixel sensors, also known as CMOS sensors (complementary MOS sensors), are generally used. However, the large quality advantage CCDs enjoyed early on has narrowed over time.

George E. Smith and Willard Boyle.

The basis for the CCD is the metal–oxide–semiconductor (MOS) structure, with MOS capacitors being the basic building blocks of a CCD, and a depleted MOS structure used as the photodetector in early CCD devices. MOS technology was originally invented by Mohamed M. Atalla and Dawon Kahng at Bell Labs in 1959.

In the late 1960s, Willard Boyle and George E. Smith at Bell Labs were researching MOS technology while working on semiconductor bubble memory. They realized that an electric charge was the analogy of the magnetic bubble and that it could be stored on a tiny MOS capacitor. As it was fairly straightforward to fabricate a series of MOS capacitors in a row, they connected a suitable voltage to them so that the charge could be stepped along from one to the next. This led to the invention of the charge-coupled device by Boyle and Smith in 1969. They conceived of the design of what they termed, in their notebook, "Charge 'Bubble' Devices".

The initial paper describing the concept in April 1970 listed possible uses as memory, a delay line, and an imaging device. The device could also be used as a shift register. The essence of the design was the ability to transfer charge along the surface of a semiconductor from one storage capacitor to the next. The concept was similar in principle to the bucket-brigade device (BBD), which was developed at Philips Research Labs during the late 1960s.

The first experimental device demonstrating the principle was a row of closely spaced metal squares on an oxidized silicon surface electrically accessed by wire bonds. It was demonstrated by Gil Amelio, Michael Francis Tompsett and George Smith in April 1970. This was the first experimental application of the CCD in image sensor technology, and used a depleted MOS structure as the photodetector. The first patent (U.S. Patent 4,085,456) on the application of CCDs to imaging was assigned to Tompsett, who filed the application in 1971.

The first working CCD made with integrated circuit technology was a simple 8-bit shift register, reported by Tompsett, Amelio and Smith in August 1970. This device had input and output circuits and was used to demonstrate its use as a shift register and as a crude eight pixel linear imaging device. Development of the device progressed at a rapid rate. By 1971, Bell researchers led by Michael Tompsett were able to capture images with simple linear devices. Several companies, including Fairchild Semiconductor, RCA and Texas Instruments, picked up on the invention and began development programs. Fairchild's effort, led by ex-Bell researcher Gil Amelio, was the first with commercial devices, and by 1974 had a linear 500-element device and a 2-D 100 x 100 pixel device. Steven Sasson, an electrical engineer working for Kodak, invented the first digital still camera using a Fairchild 100 x 100 CCD in 1975.

The interline transfer (ILT) CCD device was proposed by L. Walsh and R. Dyck at Fairchild in 1973 to reduce smear and eliminate a mechanical shutter. To further reduce smear from bright light sources, the frame-interline-transfer (FIT) CCD architecture was developed by K. Horii, T. Kuroda and T. Kunii at Matsushita (now Panasonic) in 1981.

The first KH-11 KENNEN reconnaissance satellite equipped with charge-coupled device array (800 x 800 pixels) technology for imaging was launched in December 1976. Under the leadership of Kazuo Iwama, Sony started a large development effort on CCDs involving a significant investment. Eventually, Sony managed to mass-produce CCDs for their camcorders. Before this happened, Iwama died in August 1982; subsequently, a CCD chip was placed on his tombstone to acknowledge his contribution. The first mass-produced consumer CCD video camera was released by Sony in 1983, based on a prototype developed by Yoshiaki Hagiwara in 1981.

Early CCD sensors suffered from shutter lag. This was largely resolved with the invention of the pinned photodiode (PPD). It was invented by Nobukazu Teranishi, Hiromitsu Shiraki and Yasuo Ishihara at NEC in 1980. They recognized that lag can be eliminated if the signal carriers could be transferred from the photodiode to the CCD. This led to their invention of the pinned photodiode, a photodetector structure with low lag, low noise, high quantum efficiency and low dark current. It was first publicly reported by Teranishi and Ishihara with A. Kohono, E. Oda and K. Arai in 1982, with the addition of an anti-blooming structure. The new photodetector structure invented at NEC was given the name "pinned photodiode" (PPD) by B.C. Burkey at Kodak in 1984. In 1987, the PPD began to be incorporated into most CCD devices, becoming a fixture in consumer electronic video cameras and then digital still cameras. Since then, the PPD has been used in nearly all CCD sensors and then CMOS sensors.

In January 2006, Boyle and Smith were awarded the National Academy of Engineering Charles Stark Draper Prize,and in 2009 they were awarded the Nobel Prize for Physics,for their invention of the CCD concept. Michael Tompsett was awarded the 2010 National Medal of Technology and Innovation, for pioneering work and electronic technologies including the design and development of the first CCD imagers. He was also awarded the 2012 IEEE Edison Medal for "pioneering contributions to imaging devices including CCD Imagers, cameras and thermal imagers".

Basics of Operation

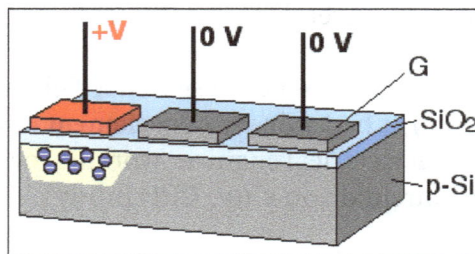

The charge packets (electrons, blue) are collected in *potential wells* (yellow) created by applying positive voltage at the gate electrodes (G). Applying positive voltage to the gate electrode in the correct sequence transfers the charge packets.

In a CCD for capturing images, there is a photoactive region (an epitaxial layer of silicon), and a transmission region made out of a shift register (the CCD, properly speaking).

An image is projected through a lens onto the capacitor array (the photoactive region), causing each capacitor to accumulate an electric charge proportional to the light intensity at that location. A one-dimensional array, used in line-scan cameras, captures a single slice of the image, whereas a two-dimensional array, used in video and still cameras, captures a two-dimensional picture corresponding to the scene projected onto the focal plane of the sensor. Once the array has been exposed to the image, a control circuit causes each capacitor to transfer its contents to its neighbor (operating as a shift register). The last capacitor in the array dumps its charge into a charge amplifier, which converts the charge into a voltage. By repeating this process, the controlling circuit converts the entire contents of the array in the semiconductor to a sequence of voltages. In a digital device, these voltages are then sampled, digitized, and usually stored in memory; in an analog device (such as an analog video camera), they are processed into a continuous analog signal (e.g. by feeding the output of the charge amplifier into a low-pass filter), which is then processed and fed out to other circuits for transmission, recording, or other processing.

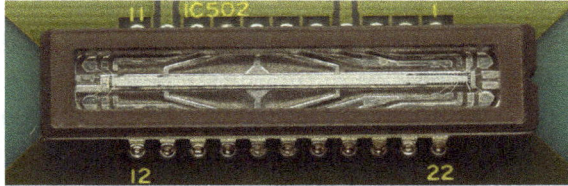
"One-dimensional" CCD image sensor from a fax machine.

Detailed Physics of Operation

Charge Generation

Before the MOS capacitors are exposed to light, they are biased into the depletion region; in n-channel CCDs, the silicon under the bias gate is slightly p-doped or intrinsic. The gate is then biased at a positive potential, above the threshold for strong inversion, which will eventually result in the creation of a n channel below the gate as in a MOS-FET. However, it takes time to reach this thermal equilibrium: up to hours in high-end scientific cameras cooled at low temperature.Initially after biasing, the holes are pushed far into the substrate, and no mobile electrons are at or near the surface; the CCD thus operates in a non-equilibrium state called deep depletion. Then, when electron–hole pairs are generated in the depletion region, they are separated by the electric field, the electrons move toward the surface, and the holes move toward the substrate. Four pair-generation processes can be identified:

- Photo-generation (up to 95% of quantum efficiency),

- Generation in the depletion region,

- Generation at the surface,

- Generation in the neutral bulk.

The last three processes are known as dark-current generation, and add noise to the image; they can limit the total usable integration time. The accumulation of electrons at or near the surface can proceed either until image integration is over and charge begins to be transferred, or thermal equilibrium is reached. In this case, the well is said to be full. The maximum capacity of each well is known as the well depth, typically about 10^5 electrons per pixel.

Design and Manufacturing

The photoactive region of a CCD is, generally, an epitaxial layer of silicon. It is lightly p doped (usually with boron) and is grown upon a substrate material, often p++. In buried-channel devices, the type of design utilized in most modern CCDs, certain areas of the surface of the silicon are ion implanted with phosphorus, giving them an n-doped designation. This region defines the channel in which the photogenerated charge packets will travel. Simon Sze details the advantages of a buried-channel device:

> This thin layer (= 0.2–0.3 micron) is fully depleted and the accumulated photogenerated charge is kept away from the surface. This structure has the advantages of higher transfer efficiency and lower dark current, from reduced surface recombination. The penalty is smaller charge capacity, by a factor of 2–3 compared to the surface-channel CCD.

The gate oxide, i.e. the capacitor dielectric, is grown on top of the epitaxial layer and substrate.

Later in the process, polysilicon gates are deposited by chemical vapor deposition, patterned with photolithography, and etched in such a way that the separately phased gates lie perpendicular to the channels. The channels are further defined by utilization of the LOCOS process to produce the channel stop region.

Channel stops are thermally grown oxides that serve to isolate the charge packets in one column from those in another. These channel stops are produced before the polysilicon gates are, as the LOCOS process utilizes a high-temperature step that would destroy the gate material. The channel stops are parallel to, and exclusive of, the channel, or "charge carrying", regions.

Channel stops often have a p+ doped region underlying them, providing a further barrier to the electrons in the charge packets.

The clocking of the gates, alternately high and low, will forward and reverse bias the diode that is provided by the buried channel (n-doped) and the epitaxial layer (p-doped). This will cause the CCD to deplete, near the p–n junction and will collect and move the charge packets beneath the gates—and within the channels—of the device.

CCD manufacturing and operation can be optimized for different uses. The above process describes a frame transfer CCD. While CCDs may be manufactured on a heavily doped p++ wafer it is also possible to manufacture a device inside p-wells that have been placed on an n-wafer. This second method, reportedly, reduces smear, dark current, and infrared and red response. This method of manufacture is used in the construction of interline-transfer devices.

Another version of CCD is called a peristaltic CCD. In a peristaltic charge-coupled device, the charge-packet transfer operation is analogous to the peristaltic contraction and dilation of the digestive system. The peristaltic CCD has an additional implant that keeps the charge away from the silicon/silicon dioxide interface and generates a large lateral electric field from one gate to the next. This provides an additional driving force to aid in transfer of the charge packets.

Architecture

The CCD image sensors can be implemented in several different architectures. The most common are full-frame, frame-transfer, and interline. The distinguishing characteristic of each of these architectures is their approach to the problem of shuttering.

In a full-frame device, all of the image area is active, and there is no electronic shutter. A mechanical shutter must be added to this type of sensor or the image smears as the device is clocked or read out.

With a frame-transfer CCD, half of the silicon area is covered by an opaque mask (typically aluminum). The image can be quickly transferred from the image area to the opaque area or storage region with acceptable smear of a few percent. That image can then be read out slowly from the storage region while a new image is integrating or exposing in the active area. Frame-transfer devices typically do not require a mechanical shutter and were a common architecture for early solid-state broadcast cameras. The downside to the frame-transfer architecture is that it requires twice the silicon real estate of an equivalent full-frame device; hence, it costs roughly twice as much.

The interline architecture extends this concept one step further and masks every other column of the image sensor for storage. In this device, only one pixel shift has to occur to transfer from image area to storage area; thus, shutter times can be less than a microsecond and smear is essentially eliminated. The advantage is not free, however, as the imaging area is now covered by opaque strips dropping the fill factor to approximately 50 percent and the effective quantum efficiency by an equivalent amount. Modern designs have addressed this deleterious characteristic by adding microlenses on the surface of the device to direct light away from the opaque regions and on the active area. Microlenses can bring the fill factor back up to 90 percent or more depending on pixel size and the overall system's optical design.

CCD from a 2.1 megapixel Argus digital camera.

CCD SONY ICX493AQA 10,14 (Gross 10,75) megapixels APS-C 1.8" 28.328mm (23.4 x 15.6 mm) from module IS-026 from digital camera SONY α (Sony Alpha) DSLR-A200 or DSLR-A300 sensor side.

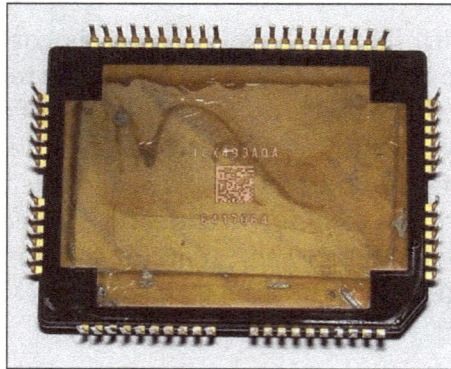

CCD SONY ICX493AQA 10,14 (Gross 10,75) megapixels APS-C 1.8" 28.328mm (23.4 x 15.6 mm) from module IS-026 from digital camera SONY α (Sony Alpha) DSLR-A200 or DSLR-A300 pins side.

The choice of architecture comes down to one of utility. If the application cannot tolerate an expensive, failure-prone, power-intensive mechanical shutter, an interline device is the right choice. Consumer snap-shot cameras have used interline devices. On the other hand, for those applications that require the best possible light collection and issues of money, power and time are less important, the full-frame device is the right choice. Astronomers tend to prefer full-frame devices. The frame-transfer falls in between and was a common choice before the fill-factor issue of interline devices was addressed. Today, frame-transfer is usually chosen when an interline architecture is not available, such as in a back-illuminated device.

CCDs containing grids of pixels are used in digital cameras, optical scanners, and video cameras as light-sensing devices. They commonly respond to 70 percent of the incident light (meaning a quantum efficiency of about 70 percent) making them far more efficient than photographic film, which captures only about 2 percent of the incident light.

CCD from a 2.1 megapixel Hewlett-Packard digital camera.

Most common types of CCDs are sensitive to near-infrared light, which allows infrared photography, night-vision devices, and zero lux (or near zero lux) video-recording/photography. For normal silicon-based detectors, the sensitivity is limited to 1.1 μm. One other consequence of their sensitivity to infrared is that infrared from remote controls often appears on CCD-based digital cameras or camcorders if they do not have infrared blockers.

Cooling reduces the array's dark current, improving the sensitivity of the CCD to low light intensities, even for ultraviolet and visible wavelengths. Professional observatories often cool their detectors with liquid nitrogen to reduce the dark current, and therefore the thermal noise, to negligible levels.

Frame Transfer CCD

A frame transfer CCD sensor.

The frame transfer CCD imager was the first imaging structure proposed for CCD Imaging by Michael Tompsett at Bell Laboratories. A frame transfer CCD is a specialized

CCD, often used in astronomy and some professional video cameras, designed for high exposure efficiency and correctness.

The normal functioning of a CCD, astronomical or otherwise, can be divided into two phases: exposure and readout. During the first phase, the CCD passively collects incoming photons, storing electrons in its cells. After the exposure time is passed, the cells are read out one line at a time. During the readout phase, cells are shifted down the entire area of the CCD. While they are shifted, they continue to collect light. Thus, if the shifting is not fast enough, errors can result from light that falls on a cell holding charge during the transfer. These errors are referred to as "vertical smear" and cause a strong light source to create a vertical line above and below its exact location. In addition, the CCD cannot be used to collect light while it is being read out. Unfortunately, a faster shifting requires a faster readout, and a faster readout can introduce errors in the cell charge measurement, leading to a higher noise level.

A frame transfer CCD solves both problems: it has a shielded, not light sensitive, area containing as many cells as the area exposed to light. Typically, this area is covered by a reflective material such as aluminium. When the exposure time is up, the cells are transferred very rapidly to the hidden area. Here, safe from any incoming light, cells can be read out at any speed one deems necessary to correctly measure the cells' charge. At the same time, the exposed part of the CCD is collecting light again, so no delay occurs between successive exposures.

The disadvantage of such a CCD is the higher cost: the cell area is basically doubled, and more complex control electronics are needed.

Intensified Charge-coupled Device

An intensified charge-coupled device (ICCD) is a CCD that is optically connected to an image intensifier that is mounted in front of the CCD.

An image intensifier includes three functional elements: a photocathode, a micro-channel plate (MCP) and a phosphor screen. These three elements are mounted one close behind the other in the mentioned sequence. The photons which are coming from the light source fall onto the photocathode, thereby generating photoelectrons. The photoelectrons are accelerated towards the MCP by an electrical control voltage, applied between photocathode and MCP. The electrons are multiplied inside of the MCP and thereafter accelerated towards the phosphor screen. The phosphor screen finally converts the multiplied electrons back to photons which are guided to the CCD by a fiber optic or a lens.

An image intensifier inherently includes a shutter functionality: If the control voltage between the photocathode and the MCP is reversed, the emitted photoelectrons are not accelerated towards the MCP but return to the photocathode. Thus, no electrons are multiplied and emitted by the MCP, no electrons are going to the phosphor screen and no light is emitted from the image intensifier. In this case no light falls onto the CCD, which means that the shutter is closed. The process of reversing the control voltage at the photocathode is called *gating* and therefore ICCDs are also called gateable CCD cameras.

Besides the extremely high sensitivity of ICCD cameras, which enable single photon detection, the gateability is one of the major advantages of the ICCD over the EMCCD cameras. The highest performing ICCD cameras enable shutter times as short as 200 picoseconds.

ICCD cameras are in general somewhat higher in price than EMCCD cameras because they need the expensive image intensifier. On the other hand, EMCCD cameras need a cooling system to cool the EMCCD chip down to temperatures around 170 K. This cooling system adds additional costs to the EMCCD camera and often yields heavy condensation problems in the application.

ICCDs are used in night vision devices and in various scientific applications.

Electron-multiplying CCD

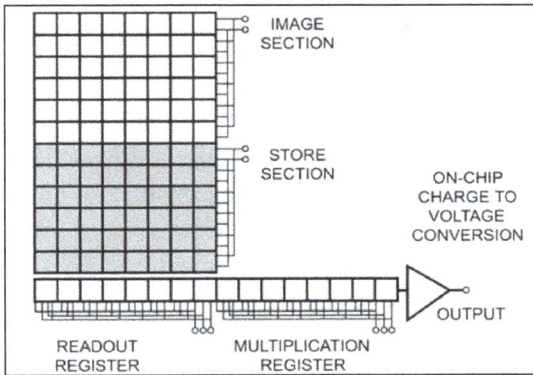

Electrons are transferred serially through the gain stages making up the multiplication register of an EMCCD. The high voltages used in these serial transfers induce the creation of additional charge carriers through impact ionisation.

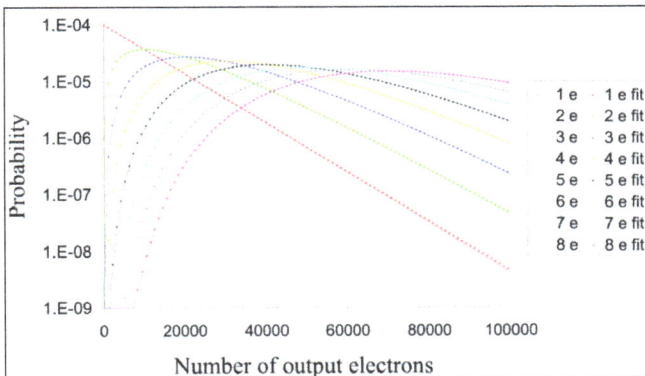

In an EMCCD there is a dispersion (variation) in the number of electrons output by the multiplication register for a given (fixed) number of input electrons (shown in the legend on the right).

The probability distribution for the number of output electrons is plotted logarithmically on the vertical axis for a simulation of a multiplication register. Also shown are results from the empirical fit equation shown on this page.

An electron-multiplying CCD (EMCCD, also known as an L3Vision CCD, a product commercialized by e2v Ltd., GB, L3CCD or Impactron CCD, a now-discontinued product offered in the past by Texas Instruments) is a charge-coupled device in which a gain register is placed between the shift register and the output amplifier. The gain register is split up into a large number of stages. In each stage, the electrons are multiplied by impact ionization in a similar way to an avalanche diode. The gain probability at every stage of the register is small ($P < 2\%$), but as the number of elements is large ($N > 500$), the overall gain can be very high ($g = (1+P)^N$), with single input electrons giving many thousands of output electrons. Reading a signal from a CCD gives a noise background, typically a few electrons. In an EMCCD, this noise is superimposed on many thousands of electrons rather than a single electron; the devices' primary advantage is thus their negligible readout noise. It is to be noted that the use of avalanche breakdown for amplification of photo charges had already been described in the U.S. Patent 3,761,744 in 1973 by George E. Smith/Bell Telephone Laboratories.

EMCCDs show a similar sensitivity to intensified CCDs (ICCDs). However, as with IC-CDs, the gain that is applied in the gain register is stochastic and the *exact* gain that has been applied to a pixel's charge is impossible to know. At high gains (> 30), this uncertainty has the same effect on the signal-to-noise ratio (SNR) as halving the quantum efficiency (QE) with respect to operation with a gain of unity. However, at very low light levels (where the quantum efficiency is most important), it can be assumed that a pixel either contains an electron — or not. This removes the noise associated with the stochastic multiplication at the risk of counting multiple electrons in the same pixel as a single electron. To avoid multiple counts in one pixel due to coincident photons in this mode of operation, high frame rates are essential. The dispersion in the gain is shown in the graph on the right. For multiplication registers with many elements and large gains it is well modelled by the equation:

$$P(n) = \frac{(n-m+1)^{m-1}}{(m-1)!\left(g-1+\dfrac{1}{m}\right)^m} \exp\left(-\frac{n-m+1}{g-1+\dfrac{1}{m}}\right) \text{ if } n \geq m$$

where P is the probability of getting n output electrons given m input electrons and a total mean multiplication register gain of g.

Because of the lower costs and better resolution, EMCCDs are capable of replacing ICCDs in many applications. ICCDs still have the advantage that they can be gated very fast and thus are useful in applications like range-gated imaging. EMCCD cameras indispensably need a cooling system — using either thermoelectric cooling or liquid nitrogen — to cool the chip down to temperatures in the range of −65 to −95 °C (−85 to −139 °F). This cooling system unfortunately adds additional costs to the EMCCD imaging system and may yield condensation problems in the application. However, high-end EMCCD cameras are equipped with a permanent hermetic vacuum system confining the chip to avoid condensation issues.

The low-light capabilities of EMCCDs find use in astronomy and biomedical research, among other fields. In particular, their low noise at high readout speeds makes them very useful for a variety of astronomical applications involving low light sources and transient events such as lucky imaging of faint stars, high speed photon counting photometry, Fabry-Pérot spectroscopy and high-resolution spectroscopy. More recently, these types of CCDs have broken into the field of biomedical research in low-light applications including small animal imaging, single-molecule imaging, Raman spectroscopy, super resolution microscopy as well as a wide variety of modern fluorescence microscopy techniques thanks to greater SNR in low-light conditions in comparison with traditional CCDs and ICCDs.

In terms of noise, commercial EMCCD cameras typically have clock-induced charge (CIC) and dark current (dependent on the extent of cooling) that together lead to an effective readout noise ranging from 0.01 to 1 electrons per pixel read. However, recent improvements in EMCCD technology have led to a new generation of cameras capable of producing significantly less CIC, higher charge transfer efficiency and an EM gain 5 times higher than what was previously available. These advances in low-light detection lead to an effective total background noise of 0.001 electrons per pixel read, a noise floor unmatched by any other low-light imaging device.

Use in Astronomy

Due to the high quantum efficiencies of charge-coupled device (CCD) (for a quantum efficiency of 100%, one count equals one photon), linearity of their outputs, ease of use compared to photographic plates, and a variety of other reasons, CCDs were very rapidly adopted by astronomers for nearly all UV-to-infrared applications.

Thermal noise and cosmic rays may alter the pixels in the CCD array. To counter such effects, astronomers take several exposures with the CCD shutter closed and opened. The average of images taken with the shutter closed is necessary to lower the random noise. Once developed, the dark frame average image is then subtracted from the open-shutter image to remove the dark current and other systematic defects (dead pixels, hot pixels, etc.) in the CCD.

The Hubble Space Telescope, in particular, has a highly developed series of steps ("data reduction pipeline") to convert the raw CCD data to useful images.

CCD cameras used in astrophotography often require sturdy mounts to cope with vibrations from wind and other sources, along with the tremendous weight of most imaging platforms. To take long exposures of galaxies and nebulae, many astronomers use a technique known as auto-guiding. Most autoguiders use a second CCD chip to monitor deviations during imaging. This chip can rapidly detect errors in tracking and command the mount motors to correct for them.

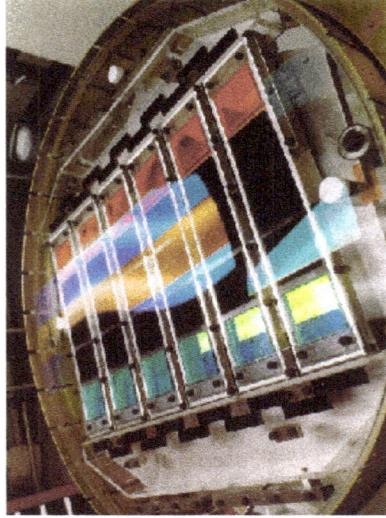

Array of 30 CCDs used on the Sloan Digital Sky Survey
telescope imaging camera, an example of "drift-scanning".

An unusual astronomical application of CCDs, called drift-scanning, uses a CCD to make a fixed telescope behave like a tracking telescope and follow the motion of the sky. The charges in the CCD are transferred and read in a direction parallel to the motion of the sky, and at the same speed. In this way, the telescope can image a larger region of the sky than its normal field of view. The Sloan Digital Sky Survey is the most famous example of this, using the technique to a survey of over a quarter of the sky.

In addition to imagers, CCDs are also used in an array of analytical instrumentation including spectrometersand interferometers.

Color Cameras

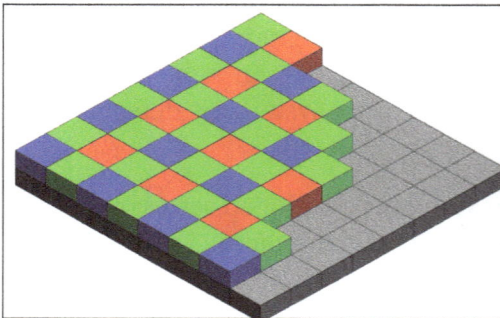

A Bayer filter on a CCD.   CCD color sensor.

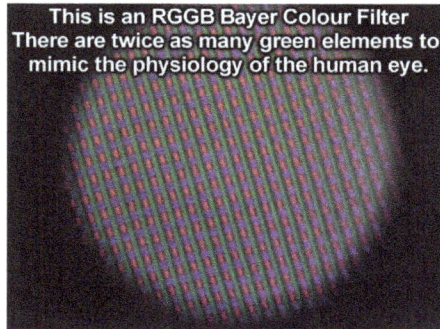

x80 microscope view of an RGGB Bayer filter on a 240 line Sony CCD PAL Camcorder CCD sensor.

Digital color cameras generally use a Bayer mask over the CCD. Each square of four pixels has one filtered red, one blue, and two green (the human eye is more sensitive to green than either red or blue). The result of this is that luminance information is collected at every pixel, but the color resolution is lower than the luminance resolution.

Better color separation can be reached by three-CCD devices (3CCD) and a dichroic beam splitter prism, that splits the image into red, green and blue components. Each of the three CCDs is arranged to respond to a particular color. Many professional video camcorders, and some semi-professional camcorders, use this technique, although developments in competing CMOS technology have made CMOS sensors, both with beam-splitters and bayer filters, increasingly popular in high-end video and digital cinema cameras. Another advantage of 3CCD over a Bayer mask device is higher quantum efficiency (and therefore higher light sensitivity for a given aperture size). This is because in a 3CCD device most of the light entering the aperture is captured by a sensor, while a Bayer mask absorbs a high proportion (about 2/3) of the light falling on each CCD pixel.

For still scenes, for instance in microscopy, the resolution of a Bayer mask device can be enhanced by microscanning technology. During the process of color co-site sampling, several frames of the scene are produced. Between acquisitions, the sensor is moved in pixel dimensions, so that each point in the visual field is acquired consecutively by elements of the mask that are sensitive to the red, green and blue components of its color. Eventually every pixel in the image has been scanned at least once in each color and the resolution of the three channels become equivalent (the resolutions of red and blue channels are quadrupled while the green channel is doubled).

Sensor Sizes

Sensors (CCD/CMOS) come in various sizes, or image sensor formats. These sizes are often referred to with an inch fraction designation such as 1/1.8″ or 2/3″ called the optical format. This measurement actually originates back in the 1950s and the time of Vidicon tubes.

Blooming

Vertical smear.

When a CCD exposure is long enough, eventually the electrons that collect in the "bins" in the brightest part of the image will overflow the bin, resulting in blooming. The structure of the CCD allows the electrons to flow more easily in one direction than another, resulting in vertical streaking.

Some anti-blooming features that can be built into a CCD reduce its sensitivity to light by using some of the pixel area for a drain structure. James M. Early developed a vertical anti-blooming drain that would not detract from the light collection area, and so did not reduce light sensitivity.

MERCURY CADMIUM TELLURIDE

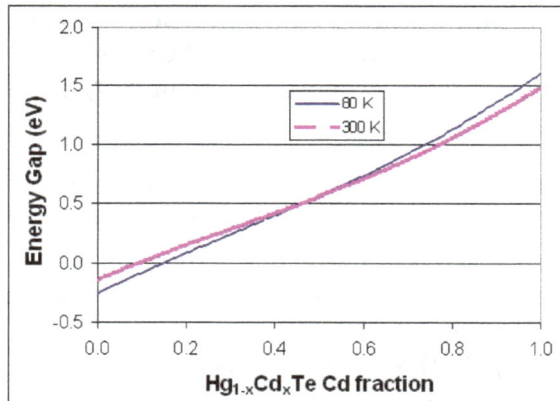

Energy gap as a function of cadmium composition.

$Hg_{1-x}Cd_xTe$ or mercury cadmium telluride (also cadmium mercury telluride, MCT, MerCad Telluride, MerCadTel, MerCaT or CMT) is an chemical compound of cadmium telluride (CdTe) and mercury telluride (HgTe) with a tunable bandgap spanning the shortwave infrared to the very long wave infrared regions. The amount of cadmium (Cd) in the alloy can be chosen so as to tune the optical absorption of the material to the desired infrared wavelength. CdTe is a semiconductor with a bandgap of approximately 1.5 electronvolts (eV) at room temperature. HgTe is a semimetal, which means that its bandgap energy is zero. Mixing these two substances allows one to obtain any bandgap between 0 and 1.5 eV.

Properties

Physical

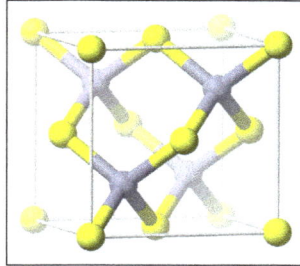

A zincblende unit cell.

$Hg_{1-x}Cd_xTe$ has a zincblende structure with two interpenetrating face-centered cubic lattices offset by $(1/4,1/4,1/4)a_o$ in the primitive cell. The cations Cd are Hg statistically mixed on the yellow sublattice while the Te anions form the grey sublattice in the image.

Electronic

The electron mobility of HgCdTe with a large Hg content is very high. Among common semiconductors used for infrared detection, only InSb and InAs surpass electron mobility of HgCdTe at room temperature. At 80 K, the electron mobility of $Hg_{0.8}Cd_{0.2}Te$ can be several hundred thousand cm²/(V·s). Electrons also have a long ballistic length at this temperature; their mean free path can be several micrometres.

The intrinsic carrier concentration is given by:

$$n_i(t,x) = (5.585 - 3.82x + (1.753 \cdot 10^{-3})t - 1.364 \cdot 10^{-3} t \cdot x) \cdot 10^{14} \cdot E_g(t,x)^{0.75} \cdot t^{1.5} \cdot e^{\frac{-E_g(t,x)\cdot q}{2\cdot k\cdot t}}$$

where k is Boltzmann's constant, q is the elementary electric charge, t is the material temperature, x is the percentage of cadmium concentration, and E_g is the bandgap given by:

$$E_g(t,x) = -0.302 + 1.93 \cdot x + (5.35 \cdot 10^{-4}) \cdot t \cdot (1 - 2 \cdot x) - 0.81 \cdot x^2 + 0.832 \cdot x^3$$

Relationship between Bandgap and Cutoff Wavelength

Using the relationship $\lambda_p = \dfrac{1.24}{E_g}$, where λ is in μm and E_g. is in electron volts, one can also obtain the cutoff wavelength as a function of x and t:

$$\lambda_p = (-0.244 + 1.556 \cdot x + (4.31 \cdot 10^{-4}) \cdot t \cdot (1 - 2 \cdot x) - 0.65 \cdot x^2 + 0.671 \cdot x^3)^{-1}$$

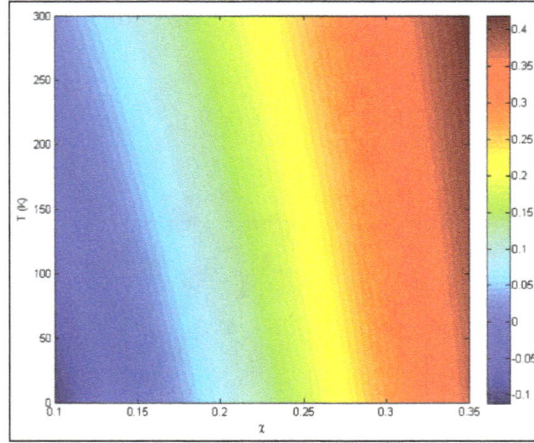

HgCdTe Bandgap in electron volts as a function of x composition and temperature.

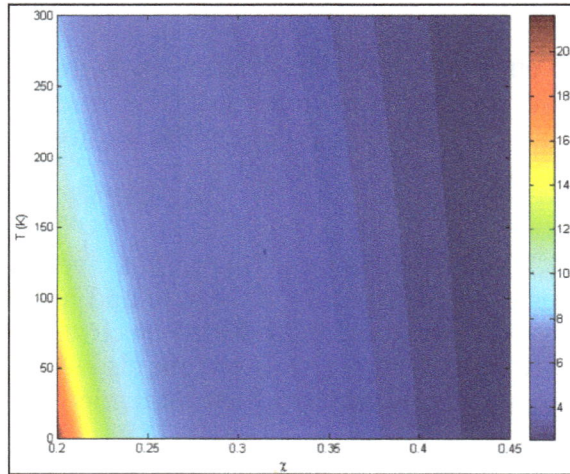

HgCdTe cutoff wavelength in μm as function of x composition and temperature.

Minority Carrier Lifetime

Auger Recombination

Two types of Auger recombination affect HgCdTe: Auger 1 and Auger 7 recombination. Auger 1 recombination involves two electrons and one hole, where an electron and a hole combine and the remaining electron receives energy equal to or greater than the band gap. Auger 7 recombination is similar to Auger 1, but involves one electron and two holes.

The Auger 1 minority carrier lifetime for intrinsic (undoped) HgCdTe is given by:

$$\tau_{Auger1}(t,x) = \frac{2.12 \cdot 10^{-14} \cdot \sqrt{E_g(t,x)} \cdot e^{\frac{q \cdot E_g(t,x)}{k \cdot t}}}{FF^2 \cdot (\frac{k \cdot t}{q})^{1.5}}$$

where FF is the overlap integral (approximately 0.221).

The Auger 1 minority carrier lifetime for doped HgCdTe is given by:

$$\tau_{Auger1_{doped}}(t,x,n) = \frac{2 \cdot \tau_{Auger1(t,x)}}{1 + (\dfrac{n}{n_i(t,x)})^2}$$

where n is the equilibrium electron concentration.

The Auger 7 minority carrier lifetime for intrinsic HgCdTe is approximately 10 times longer than the Auger 1 minority carrier lifetime:

$$\tau_{Auger7}(t,x) = 10 \cdot \tau_{Auger1}(t,x)$$

The Auger 7 minority carrier lifetime for doped HgCdTe is given by:

$$\tau_{Auger7_{doped}}(t,x,n) = \frac{2 \cdot \tau_{Auger7(t,x)}}{1 + (\dfrac{n_i(t,x)}{n})^2}$$

The total contribution of Auger 1 and Auger 7 recombination to the minority carrier lifetime is computed as:

$$\tau_{Auger}(t,x) = \frac{\tau_{Auger1}(t,x) \cdot \tau_{Auger7}(t,x)}{\tau_{Auger1}(t,x) + \tau_{Auger7}(t,x)}$$

Mechanical

HgCdTe is a soft material due to the weak bonds Hg forms with tellurium. It is a softer material than any common III-V semiconductor. The Mohs hardness of HgTe is 1.9, CdTe is 2.9 and $Hg_{0.5}Cd_{0.5}Te$ is 4. The hardness of lead salts is lower still.

Thermal

The thermal conductivity of HgCdTe is low; at low cadmium concentrations it is as low as 0.2 W·K^{-1}m^{-1}. This means that it is unsuitable for high power devices. Although infrared light-emitting diodes and lasers have been made in HgCdTe, they must be operated cold to be efficient. The specific heat capacity is 150 J·kg^{-1}K^{-1}.

Optical

HgCdTe is transparent in the infrared at photon energies below the energy gap. The refractive index is high, reaching nearly 4 for HgCdTe with high Hg content.

Infrared Detection

HgCdTe is the only common material that can detect infrared radiation in both of the accessible atmospheric windows. These are from 3 to 5 μm (the mid-wave infrared window, abbreviated MWIR) and from 8 to 12 μm (the long-wave window, LWIR). Detection in the MWIR and LWIR windows is obtained using 30% [$(Hg_{0.7}Cd_{0.3})Te$] and 20% [$(Hg_{0.8}Cd_{0.2})Te$] cadmium respectively. HgCdTe can also detect in the short wave infrared SWIR atmospheric windows of 2.2 to 2.4 μm and 1.5 to 1.8 μm.

HgCdTe is a common material in photodetectors of Fourier transform infrared spectrometers. This is because of the large spectral range of HgCdTe detectors and also the high quantum efficiency. It is also found in military field, remote sensing and infrared astronomy research. Military technology has depended on HgCdTe for night vision. In particular, the US air force makes extensive use of HgCdTe on all aircraft, and to equip airborne smart bombs. A variety of heat-seeking missiles are also equipped with HgCdTe detectors. HgCdTe detector arrays can also be found at most of the worlds major research telescopes including several satellites. Many HgCdTe detectors (such as *Hawaii* and *NICMOS* detectors) are named after the astronomical observatories or instruments for which they were originally developed.

The main limitation of LWIR HgCdTe-based detectors is that they need cooling to temperatures near that of liquid nitrogen (77K), to reduce noise due to thermally excited current carriers. MWIR HgCdTe cameras can be operated at temperatures accessible to thermoelectric coolers with a small performance penalty. Hence, HgCdTe detectors are relatively heavy compared to bolometers and require maintenance. On the other side, HgCdTe enjoys much higher speed of detection (frame rate) and is significantly more sensitive than some of its more economical competitors.

HgCdTe can be used as a heterodyne detector, in which the interference between a local source and returned laser light is detected. In this case it can detect sources such as CO_2 lasers. In heterodyne detection mode HgCdTe can be uncooled, although greater sensitivity is achieved by cooling. Photodiodes, photoconductors or photoelectromagnetic (PEM) modes can be used. A bandwidth well in excess of 1 GHz can be achieved with photodiode detectors.

The main competitors of HgCdTe are less sensitive Si-based bolometers, InSb and photon-counting superconducting tunnel junction (STJ) arrays. Quantum well infrared photodetectors (QWIP), manufactured from III-V semiconductor materials such as GaAs and AlGaAs, are another possible alternative, although their theoretical performance limits are inferior to HgCdTe arrays at comparable temperatures and they require the use of complicated reflection/diffraction gratings to overcome certain polarization exclusion effects which impact array responsivity. In the future, the primary competitor to HgCdTe detectors may emerge in the form of Quantum Dot Infrared Photodetectors (QDIP), based on either a colloidal or type-II superlattice structure. Unique 3-D quantum confinement effects, combined with the unipolar (non-exciton based photoelectric behavior) nature of quantum dots could allow comparable performance to HgCdTe at significantly higher operating temperatures. Initial laboratory work has shown promising results in this regard and QDIPs may be one of the first significant nanotechnology products to emerge.

In HgCdTe, detection occurs when an infrared photon of sufficient energy kicks an electron from the valence band to the conduction band. Such an electron is collected by a suitable external readout integrated circuits (ROIC) and transformed into an electric signal. The physical mating of the HgCdTe detector array to the ROIC is often referred to as a "focal plane array".

In contrast, in a bolometer, light heats up a tiny piece of material. The temperature change of the bolometer results in a change in resistance which is measured and transformed into an electric signal.

Mercury zinc telluride has better chemical, thermal, and mechanical stability characteristics than HgCdTe. It has a steeper change of energy gap with mercury composition than HgCdTe, making compositional control harder.

HgCdTe Growth Techniques

Bulk Crystal Growth

The first large scale growth method was bulk recrystallization of a liquid melt. This was the main growth method from the late 1950s to the early 1970s.

Epitaxial Growth

Highly pure and crystalline HgCdTe is fabricated by epitaxy on either CdTe or CdZnTe substrates. CdZnTe is a compound semiconductor, the lattice parameter of which can be exactly matched to that of HgCdTe. This eliminates most defects from the epilayer of HgCdTe. CdTe was developed as an alternative substrate in the '90s. It is not lattice-matched to HgCdTe, but is much cheaper, as it can be grown by epitaxy on silicon (Si) or germanium (Ge) substrates.

Liquid phase epitaxy (LPE), in which a substrate is repeatedly dipped into a liquid melt, gives the best results in terms of crystalline quality, and is still a common technique of choice for industrial production.

In recent years, molecular beam epitaxy (MBE) has become widespread because of its ability to stack up layers of different alloy composition. This allows simultaneous detection at several wavelengths. Furthermore, MBE, and also MOVPE, allow growth on large area substrates such as CdTe on Si or Ge, whereas LPE does not allow such substrates to be used.

Toxicity

The advancement of crystal growth technology has proceeded deliberately and steadily for four decades in spite of the high vapor pressure of Hg at the melting point of HgCdTe and the known toxicity of the material.

LIGHT-EMITTING DIODE

A light-emitting diode (LED) is a semiconductor light source that emits light when current flows through it. Electrons in the semiconductor recombine with electron holes, releasing energy in the form of photons. The color of the light (corresponding to the energy of the photons) is determined by the energy required for electrons to cross the band gap of the semiconductor. White light is obtained by using multiple semiconductors or a layer of light-emitting phosphor on the semiconductor device.

Appearing as practical electronic components in 1962, the earliest LEDs emitted low-intensity infrared light. Infrared LEDs are used in remote-control circuits, such as those used with a wide variety of consumer electronics. The first visible-light LEDs were of low intensity and limited to red. Modern LEDs are available across the visible, ultraviolet, and infrared wavelengths, with high light output.

Early LEDs were often used as indicator lamps, replacing small incandescent bulbs, and in seven-segment displays. Recent developments have produced high-output white light LEDs suitable for room and outdoor area lighting. LEDs have led to new displays and sensors, while their high switching rates are useful in advanced communications technology.

LEDs have many advantages over incandescent light sources, including lower energy consumption, longer lifetime, improved physical robustness, smaller size, and faster switching. LEDs are used in applications as diverse as aviation lighting, automotive headlamps, advertising, general lighting, traffic signals, camera flashes, lighted wallpaper, plant growing light, and medical devices.

Unlike a laser, the light emitted from an LED is neither spectrally coherent nor even highly monochromatic. However, its spectrum is sufficiently narrow that it appears to the human eye as a pure (saturated) color. Nor, unlike most lasers, is its radiation spatially coherent, so that it cannot approach the very high brightnesses characteristic of lasers.

Colors

By selection of different semiconductor materials, single-color LEDs can be made that emit light in a narrow band of wavelengths from near-infrared through the visible spectrum and into the ultraviolet range. As the wavelengths become shorter, because of the larger band gap of these semiconductors, the operating voltage of the LED increases.

Blue and Ultraviolet

Blue LEDs have an active region consisting of one or more InGaN quantum wells sandwiched between thicker layers of GaN, called cladding layers. By varying the relative In/Ga fraction in the InGaN quantum wells, the light emission can in theory be varied from violet to amber.

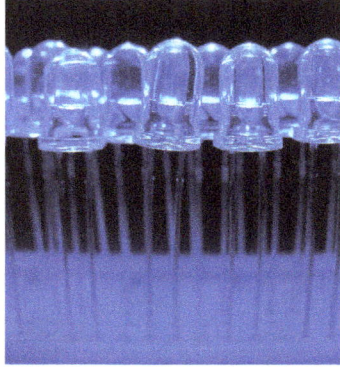

Blue LEDs.

Aluminium gallium nitride (AlGaN) of varying Al/Ga fraction can be used to manufacture the cladding and quantum well layers for ultraviolet LEDs, but these devices have not yet reached the level of efficiency and technological maturity of InGaN/GaN blue/green devices. If un-alloyed GaN is used in this case to form the active quantum well layers, the device emits near-ultraviolet light with a peak wavelength centred around 365 nm. Green LEDs manufactured from the InGaN/GaN system are far more efficient and brighter than green LEDs produced with non-nitride material systems, but practical devices still exhibit efficiency too low for high-brightness applications.

With AlGaN and AlGaInN, even shorter wavelengths are achievable. Near-UV emitters at wavelengths around 360–395 nm are already cheap and often encountered, for example, as black light lamp replacements for inspection of anti-counterfeiting UV watermarks in documents and bank notes, and for UV curing. While substantially more expensive, shorter-wavelength diodes are commercially available for wavelengths down to 240 nm. As the photosensitivity of microorganisms approximately matches the absorption spectrum of DNA, with a peak at about 260 nm, UV LED emitting at 250–270 nm are expected in prospective disinfection and sterilization devices. Recent research has shown that commercially available UVA LEDs (365 nm) are already effective disinfection and sterilization devices.UV-C wavelengths were obtained in laboratories using aluminium nitride (210 nm),boron nitride (215 nm) and diamond (235 nm).

White

There are two primary ways of producing white light-emitting diodes. One is to use individual LEDs that emit three primary colors—red, green and blue—and then mix all the colors to form white light. The other is to use a phosphor material to convert monochromatic light from a blue or UV LED to broad-spectrum white light, similar to a fluorescent lamp. The yellow phosphor is cerium-doped YAG crystals suspended in the package or coated on the LED. This YAG phosphor causes white LEDs to look yellow when off.

The 'whiteness' of the light produced is engineered to suit the human eye. Because of metamerism, it is possible to have quite different spectra that appear white. However, the appearance of objects illuminated by that light may vary as the spectrum varies. This is the issue of color rendition, quite separate from color temperature. An orange or cyan object could appear with the wrong color and much darker as the LED or phosphor does not emit the wavelength it reflects. The best color rendition LEDs use a mix of phosphors, resulting in less efficiency but better color rendering.

RGB Systems

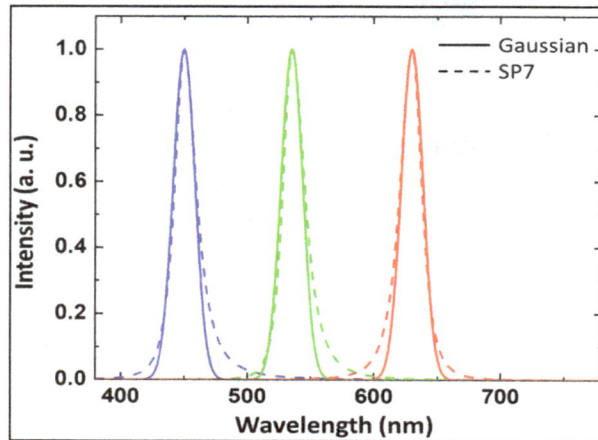

Combined spectral curves for blue, yellow-green, and high-brightness red solid-state semiconductor LEDs. FWHM spectral bandwidth is approximately 24–27 nm for all three colors.

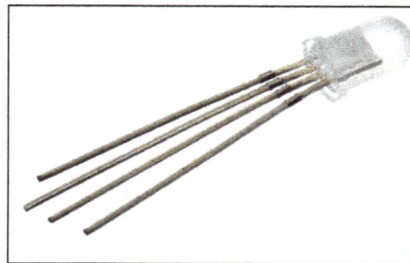

RGB LED.

Mixing red, green, and blue sources to produce white light needs electronic circuits to control the blending of the colors. Since LEDs have slightly different emission patterns, the color balance may change depending on the angle of view, even if the RGB sources are in a single package, so RGB diodes are seldom used to produce white lighting. Nonetheless, this method has many applications because of the flexibility of mixing different colors,and in principle, this mechanism also has higher quantum efficiency in producing white light.

There are several types of multicolor white LEDs: di-, tri-, and tetrachromatic white LEDs. Several key factors that play among these different methods include color stability, color rendering capability, and luminous efficacy. Often, higher efficiency means lower color rendering, presenting a trade-off between the luminous efficacy and color rendering. For example, the dichromatic white LEDs have the best luminous efficacy (120 lm/W), but the lowest color rendering capability. However, although tetrachromatic white LEDs have excellent color rendering capability, they often have poor luminous efficacy. Trichromatic white LEDs are in between, having both good luminous efficacy (>70 lm/W) and fair color rendering capability.

One of the challenges is the development of more efficient green LEDs. The theoretical maximum for green LEDs is 683 lumens per watt but as of 2010 few green LEDs exceed even 100 lumens per watt. The blue and red LEDs approach their theoretical limits.

Multicolor LEDs also offer a new means to form light of different colors. Most perceivable colors can be formed by mixing different amounts of three primary colors. This allows precise dynamic color control. However, this type of LED's emission power decays exponentially with rising temperature, resulting in a substantial change in color stability. Such problems inhibit industrial use. Multicolor LEDs without phosphors cannot provide good color rendering because each LED is a narrowband source. LEDs without phosphor, while a poorer solution for general lighting, are the best solution for displays, either backlight of LCD, or direct LED based pixels.

Dimming a multicolor LED source to match the characteristics of incandescent lamps is difficult because manufacturing variations, age, and temperature change the actual color value output. To emulate the appearance of dimming incandescent lamps may require a feedback system with color sensor to actively monitor and control the color.

Phosphor-based LEDs

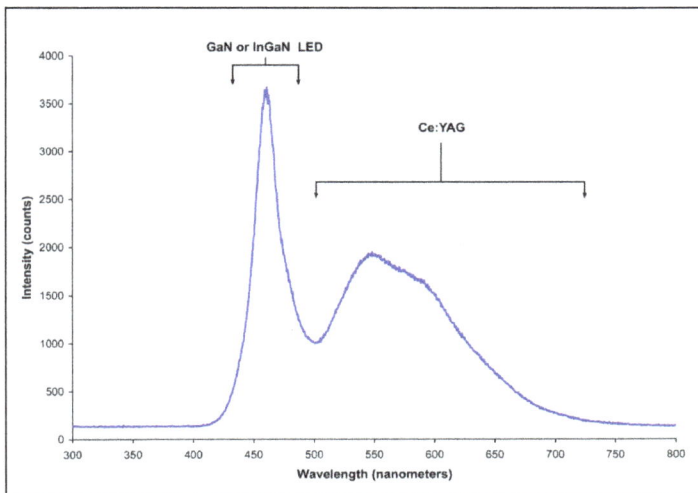

Spectrum of a white LED showing blue light directly emitted by the GaN-based LED (peak at about 465 nm) and the more broadband Stokes-shifted light emitted by the Ce^{3+}:YAG phosphor, which emits at roughly 500–700 nm.

This method involves coating LEDs of one color (mostly blue LEDs made of InGaN) with phosphors of different colors to form white light; the resultant LEDs are called phosphor-based or phosphor-converted white LEDs (pcLEDs). A fraction of the blue light undergoes the Stokes shift, which transforms it from shorter wavelengths to longer. Depending on the original LED's color, various color phosphors are used. Using several phosphor layers of distinct colors broadens the emitted spectrum, effectively raising the color rendering index (CRI).

Phosphor-based LEDs have efficiency losses due to heat loss from the Stokes shift and also other phosphor-related issues. Their luminous efficacies compared to normal LEDs depend on the spectral distribution of the resultant light output and the original wavelength of the LED itself. For example, the luminous efficacy of a typical YAG yellow phosphor based white LED ranges from 3 to 5 times the luminous efficacy of the original blue LED because of the human eye's greater sensitivity to yellow than to blue (as modeled in the luminosity function). Due to the simplicity of manufacturing, the phosphor method is still the most popular method for making high-intensity white LEDs. The design and production of a light source or light fixture using a monochrome emitter with phosphor conversion is simpler and cheaper than a complex RGB system, and the majority of high-intensity white LEDs presently on the market are manufactured using phosphor light conversion.

Among the challenges being faced to improve the efficiency of LED-based white light sources is the development of more efficient phosphors. As of 2010, the most efficient yellow phosphor is still the YAG phosphor, with less than 10% Stokes shift loss. Losses attributable to internal optical losses due to re-absorption in the LED chip and in the LED packaging itself account typically for another 10% to 30% of efficiency loss. Currently, in the area of phosphor LED development, much effort is being spent on optimizing these devices to higher light output and higher operation temperatures. For instance, the efficiency can be raised by adapting better package design or by using a more suitable type of phosphor. Conformal coating process is frequently used to address the issue of varying phosphor thickness.

Some phosphor-based white LEDs encapsulate InGaN blue LEDs inside phosphor-coated epoxy. Alternatively, the LED might be paired with a remote phosphor, a preformed polycarbonate piece coated with the phosphor material. Remote phosphors provide more diffuse light, which is desirable for many applications. Remote phosphor designs are also more tolerant of variations in the LED emissions spectrum. A common yellow phosphor material is cerium-doped yttrium aluminium garnet (Ce^{3+}:YAG).

White LEDs can also be made by coating near-ultraviolet (NUV) LEDs with a mixture of high-efficiency europium-based phosphors that emit red and blue, plus copper and aluminium-doped zinc sulfide (ZnS:Cu, Al) that emits green. This is a method analogous to the way fluorescent lamps work. This method is less efficient than blue LEDs with YAG:Ce phosphor, as the Stokes shift is larger, so more energy is converted to heat, but yields light with better spectral characteristics, which render color better. Due to the higher radiative output of the ultraviolet LEDs than of the blue ones, both methods offer comparable brightness. A concern is that UV light may leak from a malfunctioning light source and cause harm to human eyes or skin.

Other White LEDs

Another method used to produce experimental white light LEDs used no phosphors at all and was based on homoepitaxially grown zinc selenide (ZnSe) on a ZnSe substrate that simultaneously emitted blue light from its active region and yellow light from the substrate.

A new style of wafers composed of gallium-nitride-on-silicon (GaN-on-Si) is being used to produce white LEDs using 200-mm silicon wafers. This avoids the typical costly sapphire substrate in relatively small 100- or 150-mm wafer sizes.The sapphire apparatus must be coupled with a mirror-like collector to reflect light that would otherwise be wasted. It is predicted that by 2020, 40% of all GaN LEDs will be made with GaN-on-Si. Manufacturing large sapphire material is difficult, while large silicon material is cheaper and more abundant. LED companies shifting from using sapphire to silicon should be a minimal investment.

Types

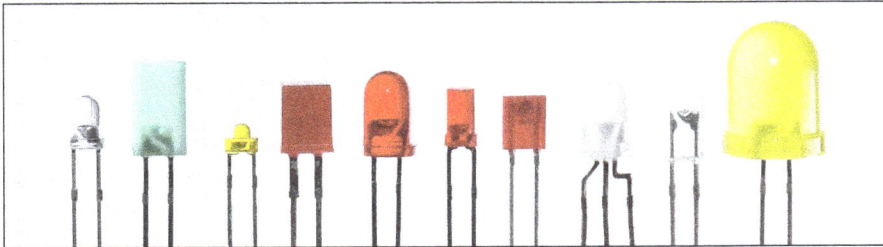

LEDs are produced in a variety of shapes and sizes. The color of the plastic lens is often the same as the actual color of light emitted, but not always.

For instance, purple plastic is often used for infrared LEDs, and most blue devices have colorless housings. Modern high-power LEDs such as those used for lighting and backlighting are generally found in surface-mount technology (SMT) packages.

LEDs are made in different packages for different applications. A single or a few LED junctions may be packed in one miniature device for use as an indicator or pilot lamp. An LED array may include controlling circuits within the same package, which may range from a simple resistor, blinking or color changing control, or an addressable controller for RGB devices. Higher-powered white-emitting devices will be mounted on heat sinks and will be used for illumination. Alphanumeric displays in dot matrix or bar formats are widely available. Special packages permit connection of LEDs to optical fibers for high-speed data communication links.

Miniature

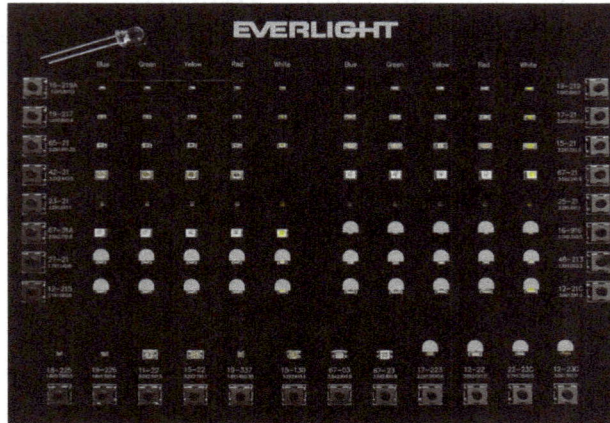

Photo of miniature surface mount LEDs in most common sizes. They can be much smaller than a traditional 5 mm lamp type LED, shown on the upper left corner.

Very small (1.6x1.6x0.35 mm) red, green, and blue surface mount miniature LED package with gold wire bonding details.

These are mostly single-die LEDs used as indicators, and they come in various sizes from 2 mm to 8 mm, through-hole and surface mount packages. Typical current ratings range from around 1 mA to above 20 mA. Multiple LED dies attached to a flexible backing tape form an LED strip light.

Common package shapes include round, with a domed or flat top, rectangular with a flat top (as used in bar-graph displays), and triangular or square with a flat top. The encapsulation may also be clear or tinted to improve contrast and viewing angle. Infrared devices may have a black tint to block visible light while passing infrared radiation.

Ultra-high-output LEDs are designed for viewing in direct sunlight. 5 V and 12 V LEDs are ordinary miniature LEDs that have a series resistor for direct connection to a 5 V or 12 V supply.

High-power

High-power light-emitting diodes attached to an LED star base.

High-power LEDs (HP-LEDs) or high-output LEDs (HO-LEDs) can be driven at currents from hundreds of mA to more than an ampere, compared with the tens of mA for other LEDs. Some can emit over a thousand lumens. LED power densities up to 300 W/cm² have been achieved. Since overheating is destructive, the HP-LEDs must be mounted on a heat sink to allow for heat dissipation. If the heat from an HP-LED is not removed, the device fails in seconds. One HP-LED can often replace an incandescent bulb in a flashlight, or be set in an array to form a powerful LED lamp.

Some well-known HP-LEDs in this category are the Nichia 19 series, Lumileds Rebel Led, Osram Opto Semiconductors Golden Dragon, and Cree X-lamp. As of September 2009, some HP-LEDs manufactured by Cree now exceed 105 lm/W.

Examples for Haitz's law—which predicts an exponential rise in light output and efficacy of LEDs over time—are the CREE XP-G series LED, which achieved 105 lm/W in 2009and the Nichia 19 series with a typical efficacy of 140 lm/W, released in 2010.

AC-driven

LEDs developed by Seoul Semiconductor can operate on AC power without a DC converter. For each half-cycle, part of the LED emits light and part is dark, and this is reversed during the next half-cycle. The efficacy of this type of HP-LED is typically 40 lm/W.A large number of LED elements in series may be able to operate directly from line voltage. In 2009, Seoul Semiconductor released a high DC voltage LED, named as 'Acrich MJT', capable of being driven from AC power with a simple controlling circuit. The low-power dissipation of these LEDs affords them more flexibility than the original AC LED design.

Application-specific Variations

Flashing

Flashing LEDs are used as attention seeking indicators without requiring external electronics. Flashing LEDs resemble standard LEDs but they contain an integrated multivibrator circuit that causes the LED to flash with a typical period of one second. In diffused lens LEDs, this circuit is visible as a small black dot. Most flashing LEDs emit light of one color, but more sophisticated devices can flash between multiple colors and even fade through a color sequence using RGB color mixing.

Bi-color

Bi-color LEDs contain two different LED emitters in one case. There are two types of these. One type consists of two dies connected to the same two leads antiparallel to each other. Current flow in one direction emits one color, and current in the opposite direction emits the other color. The other type consists of two dies with separate leads for both dies and another lead for common anode or cathode so that they can be controlled independently. The most common bi-color combination is red/traditional green, however, other available combinations include amber/traditional green, red/pure green, red/blue, and blue/pure green.

RGB Tri-color

Tri-color LEDs contain three different LED emitters in one case. Each emitter is connected to a separate lead so they can be controlled independently. A four-lead arrangement is typical with one common lead (anode or cathode) and an additional lead for each color. Others, however, have only two leads (positive and negative) and have a built-in electronic controller.

RGB-SMD-LED.

RGB LEDs consist of one red, one green, and one blue LED. By independently adjusting each of the three, RGB LEDs are capable of producing a wide color gamut. Unlike dedicated-color LEDs, however, these do not produce pure wavelengths. Modules may not be optimized for smooth color mixing.

Decorative-multicolor

Decorative-multicolor LEDs incorporate several emitters of different colors supplied by only two lead-out wires. Colors are switched internally by varying the supply voltage.

Alphanumeric

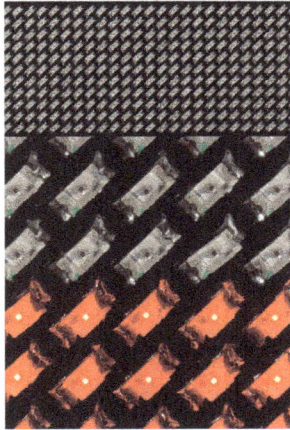

Composite image of an 11 × 44 LED matrix lapel name tag display using 1608/0603-type SMD LEDs. Top: A little over half of the 21x86 mm display. Center: Close-up of LEDs in ambient light. Bottom: LEDs in their own red light.

Alphanumeric LEDs are available in seven-segment, starburst, and dot-matrix format. Seven-segment displays handle all numbers and a limited set of letters. Starburst displays can display all letters. Dot-matrix displays typically use 5x7 pixels per character. Seven-segment LED displays were in widespread use in the 1970s and 1980s, but rising use of liquid crystal displays, with their lower power needs and greater display flexibility, has reduced the popularity of numeric and alphanumeric LED displays.

Digital RGB

Digital RGB addressable LEDs contain their own "smart" control electronics. In addition to power and ground, these provide connections for data-in, data-out, and sometimes a clock or strobe signal. These are connected in a daisy chain. Data sent to the first LED of the chain can control the brightness and color of each LED independently of the others. They are used where a combination of maximum control and minimum visible electronics are needed such as strings for Christmas and LED matrices. Some even have refresh rates in the kHz range, allowing for basic video applications. These devices are known by their part number (WS2812 being common) or a brand name such as NeoPixel.

Filament

An LED filament consists of multiple LED chips connected in series on a common longitudinal substrate that forms a thin rod reminiscent of a traditional incandescent filament. These are being used as a low-cost decorative alternative for traditional light bulbs that are being phased out in many countries. The filaments use a rather high voltage, allowing them to work efficiently with mains voltages. Often a simple rectifier and capacitive current limiting are employed to create a low-cost replacement for a traditional light bulb without the complexity of the low voltage, high current converter that single die LEDs need. Usually, they are packaged in bulb similar to the lamps they were designed to replace, and filled with inert gas to remove heat efficiently.

Chip-on-board Arrays

Surface-mounted LEDs are frequently produced in chip on board (COB) arrays, allowing better heat dissipation than with a single LED of comparable luminous output.The LEDs can be arranged around a cylinder, and are called "corn cob lights" because of the rows of yellow LEDs.

Considerations for Use

Power Sources

The current in an LED or other diodes rises exponentially with the applied voltage, so a small change in voltage can cause a large change in current. Current through the LED must be regulated by an external circuit such as a constant current source to prevent damage. Since most common power supplies are (nearly) constant-voltage sources, LED fixtures must include a power converter, or at least a current-limiting resistor. In some applications, the internal resistance of small batteries is sufficient to keep current within the LED rating.

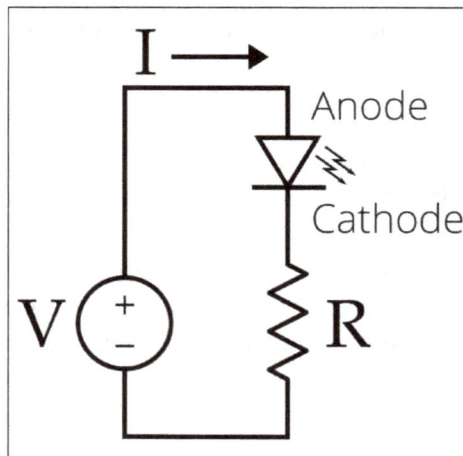

Simple LED circuit with resistor for current limiting.

Electrical Polarity

An LED will light only when voltage is applied in the forward direction of the diode. No current flows and no light is emitted if voltage is applied in the reverse direction. If the reverse voltage exceeds the breakdown voltage, a large current flows and the LED will be damaged. If the reverse current is sufficiently limited to avoid damage, the reverse-conducting LED is a useful noise diode.

Safety and Health

Certain blue LEDs and cool-white LEDs can exceed safe limits of the so-called blue-light hazard as defined in eye safety specifications such as "ANSI/IESNA RP-27.1–05: Recommended Practice for Photobiological Safety for Lamp and Lamp Systems". One study showed no evidence of a risk in normal use at domestic illuminance, and that caution is only needed for particular occupational situations or for specific populations.

While LEDs have the advantage over fluorescent lamps, in that they do not contain mercury, they may contain other hazardous metals such as lead and arsenic.

In 2016 the American Medical Association (AMA) issued a statement concerning the possible adverse influence of blueish street lighting on the sleep-wake cycle of city-dwellers. Industry critics claim exposure levels are not high enough to have a noticeable effect.

Advantages

- Efficiency: LEDs emit more lumens per watt than incandescent light bulbs. The efficiency of LED lighting fixtures is not affected by shape and size, unlike fluorescent light bulbs or tubes.

- Color: LEDs can emit light of an intended color without using any color filters as traditional lighting methods need. This is more efficient and can lower initial costs.

- Size: LEDs can be very small (smaller than 2 mm^2) and are easily attached to printed circuit boards.

- Warmup time: LEDs light up very quickly. A typical red indicator LED achieves full brightness in under a microsecond. LEDs used in communications devices can have even faster response times.

- Cycling: LEDs are ideal for uses subject to frequent on-off cycling, unlike incandescent and fluorescent lamps that fail faster when cycled often, or high-intensity discharge lamps (HID lamps) that require a long time before restarting.

- Dimming: LEDs can very easily be dimmed either by pulse-width modulation or lowering the forward current. This pulse-width modulation is why LED lights, particularly headlights on cars, when viewed on camera or by some people, seem to flash or flicker. This is a type of stroboscopic effect.

- Cool light: In contrast to most light sources, LEDs radiate very little heat in the form of IR that can cause damage to sensitive objects or fabrics. Wasted energy is dispersed as heat through the base of the LED.

- Slow failure: LEDs mainly fail by dimming over time, rather than the abrupt failure of incandescent bulbs.

- Lifetime: LEDs can have a relatively long useful life. One report estimates 35,000 to 50,000 hours of useful life, though time to complete failure may be shorter or longer. Fluorescent tubes typically are rated at about 10,000 to 25,000 hours, depending partly on the conditions of use, and incandescent light bulbs at 1,000 to 2,000 hours. Several DOE demonstrations have shown that reduced maintenance costs from this extended lifetime, rather than energy savings, is the primary factor in determining the payback period for an LED product.

- Shock resistance: LEDs, being solid-state components, are difficult to damage with external shock, unlike fluorescent and incandescent bulbs, which are fragile.

- Focus: The solid package of the LED can be designed to focus its light. Incandescent and fluorescent sources often require an external reflector to collect light and direct it in a usable manner. For larger LED packages total internal reflection (TIR) lenses are often used to the same effect. However, when large quantities of light are needed many light sources are usually deployed, which are difficult to focus or collimate towards the same target.

Disadvantages

- Temperature dependence: LED performance largely depends on the ambient temperature of the operating environment – or thermal management properties. Overdriving an LED in high ambient temperatures may result in overheating the LED package, eventually leading to device failure. An adequate heat sink is needed to maintain long life. This is especially important in automotive, medical, and military uses where devices must operate over a wide range of temperatures, which require low failure rates. Toshiba has produced LEDs with an operating temperature range of −40 to 100 °C, which suits the LEDs for both indoor and outdoor use in applications such as lamps, ceiling lighting, street lights, and floodlights.

- Voltage sensitivity: LEDs must be supplied with a voltage above their threshold voltage and a current below their rating. Current and lifetime change greatly with a small change in applied voltage. They thus require a current-regulated supply (usually just a series resistor for indicator LEDs).

- Color rendition: Most cool-white LEDs have spectra that differ significantly from a black body radiator like the sun or an incandescent light. The spike at 460 nm and dip at 500 nm can make the color of objects appear differently under cool-white LED illumination than sunlight or incandescent sources, due to metamerism,red surfaces being rendered particularly poorly by typical phosphor-based cool-white LEDs. The same is true with green surfaces.

- Area light source: Single LEDs do not approximate a point source of light giving a spherical light distribution, but rather a lambertian distribution. So, LEDs are difficult to apply to uses needing a spherical light field; however, different fields of light can be manipulated by the application of different optics or "lenses". LEDs cannot provide divergence below a few degrees.

- Light pollution: Because white LEDs emit more short wavelength light than sources such as high-pressure sodium vapor lamps, the increased blue and green sensitivity of scotopic vision means that white LEDs used in outdoor lighting cause substantially more sky glow.

- Efficiency droop: The efficiency of LEDs decreases as the electric current increases. Heating also increases with higher currents, which compromises LED lifetime. These effects put practical limits on the current through an LED in high power applications.

- Impact on insects: LEDs are much more attractive to insects than sodium-vapor lights, so much so that there has been speculative concern about the possibility of disruption to food webs.

- Use in winter conditions: Since they do not give off much heat in comparison to incandescent lights, LED lights used for traffic control can have snow obscuring them, leading to accidents.

- Thermal runaway: Parallel strings of LEDs will not share current evenly due to the manufacturing tolerance in their forward voltage. Running two or more strings from a single current source will likely result in LED failure as the devices warm up. A circuit is required to ensure even distribution of current between parallel strands.

Applications

Daytime running light LEDs of an automobile.

LED uses fall into four major categories:

- Visual signals where light goes more or less directly from the source to the human eye, to convey a message or meaning.

- Illumination where light is reflected from objects to give visual response of these objects.

- Measuring and interacting with processes involving no human vision.

- Narrow band light sensors where LEDs operate in a reverse-bias mode and respond to incident light, instead of emitting light.

Indicators and Signs

The low energy consumption, low maintenance and small size of LEDs has led to uses as status indicators and displays on a variety of equipment and installations. Large-area LED displays are used as stadium displays, dynamic decorative displays, and dynamic message signs on freeways. Thin, lightweight message displays are used at airports and railway stations, and as destination displays for trains, buses, trams, and ferries.

Red and green LED traffic signals.

One-color light is well suited for traffic lights and signals, exit signs, emergency vehicle lighting, ships' navigation lights, and LED-based Christmas lights.

Because of their long life, fast switching times, and visibility in broad daylight due to their high output and focus, LEDs have been used in automotive brake lights and turn signals. The use in brakes improves safety, due to a great reduction in the time needed to light fully, or faster rise time, about 0.1 second faster than an incandescent bulb. This gives drivers behind more time to react. In a dual intensity circuit (rear markers and brakes) if the LEDs are not pulsed at a fast enough frequency, they can create a phantom array, where ghost images of the LED appear if the eyes quickly scan across the array. White LED headlamps are beginning to appear. Using LEDs has styling advantages because LEDs can form much thinner lights than incandescent lamps with parabolic reflectors.

Due to the relative cheapness of low output LEDs, they are also used in many temporary uses such as glowsticks, throwies, and the photonic textile Lumalive. Artists have also used LEDs for LED art.

Lighting

With the development of high-efficiency and high-power LEDs, it has become possible to use LEDs in lighting and illumination. To encourage the shift to LED lamps and other high-efficiency lighting, in 2008 the US Department of Energy created the L Prize competition. The Philips Lighting North America LED bulb won the first competition on August 3, 2011, after successfully completing 18 months of intensive field, lab, and product testing.

Efficient lighting is needed for sustainable architecture. As of 2011, some LED bulbs provide up to 150 lm/W and even inexpensive low-end models typically exceed 50 lm/W, so that a 6-watt LED could achieve the same results as a standard 40-watt incandescent bulb. Displacing less effective sources such as incandescent lamps and fluorescent lighting reduces electrical energy consumption and its associated emissions.

LEDs are used as street lights and in architectural lighting. The mechanical robustness and long lifetime are used in automotive lighting on cars, motorcycles, and bicycle lights. LED street lights are employed on poles and in parking garages. In 2007, the Italian village of Torraca was the first place to convert its street lighting to LEDs.

Cabin lighting on recent Airbus and Boeing jetliners uses LED lighting. LEDs are also being used in airport and heliport lighting. LED airport fixtures currently include medium-intensity runway lights, runway centerline lights, taxiway centerline and edge lights, guidance signs, and obstruction lighting.

LEDs are also used as a light source for DLP projectors, and to backlight LCD televisions (referred to as LED TVs) and laptop displays. RGB LEDs raise the color gamut by as much as 45%. Screens for TV and computer displays can be made thinner using LEDs for backlighting.

The lower heat radiation compared with incandescent lamps makes LEDs ideal for stage lights, where banks of RGB LEDs can easily change color and decrease heating from traditional stage lighting. In medical lighting, infrared heat radiation can be harmful. In energy conservation, the lower heat output of LEDs also reduces demand on air conditioning systems.

LEDs are small, durable and need little power, so they are used in handheld devices such as flashlights. LED strobe lights or camera flashes operate at a safe, low voltage, instead of the 250+ volts commonly found in xenon flashlamp-based lighting. This is especially useful in cameras on mobile phones, where space is at a premium and bulky voltage-raising circuitry is undesirable.

LEDs are used for infrared illumination in night vision uses including security cameras. A ring of LEDs around a video camera, aimed forward into a retroreflective background, allows chroma keying in video productions.

LED for miners, to increase visibility inside mines.

Los Angeles Vincent Thomas Bridge illuminated with blue LEDs.

LEDs are used in mining operations, as cap lamps to provide light for miners. Research has been done to improve LEDs for mining, to reduce glare and to increase illumination, reducing risk of injury to the miners.

LEDs are increasingly finding uses in medical and educational applications, for example as mood enhancement,and new technologies such as AmBX, exploiting LED versatility. NASA has even sponsored research for the use of LEDs to promote health for astronauts.

Data Communication and other Signalling

Light can be used to transmit data and analog signals. For example, lighting white

LEDs can be used in systems assisting people to navigate in closed spaces while searching necessary rooms or objects.

Assistive listening devices in many theaters and similar spaces use arrays of infrared LEDs to send sound to listeners' receivers. Light-emitting diodes (as well as semiconductor lasers) are used to send data over many types of fiber optic cable, from digital audio over TOSLINK cables to the very high bandwidth fiber links that form the Internet backbone. For some time, computers were commonly equipped with IrDA interfaces, which allowed them to send and receive data to nearby machines via infrared.

Because LEDs can cycle on and off millions of times per second, very high data bandwidth can be achieved.

Machine Vision Systems

Machine vision systems often require bright and homogeneous illumination, so features of interest are easier to process. LEDs are often used.

Barcode scanners are the most common example of machine vision applications, and many of those scanners use red LEDs instead of lasers. Optical computer mice use LEDs as a light source for the miniature camera within the mouse.

LEDs are useful for machine vision because they provide a compact, reliable source of light. LED lamps can be turned on and off to suit the needs of the vision system, and the shape of the beam produced can be tailored to match the systems's requirements.

Biological Detection

UV induced fluorescence is one of the most robust techniques used for rapid real time detection of biological aerosols. The first UV sensors were lasers lacking in-field-use practicality. In order to address this, DARPA incorporated SUVOS technology to create a low cost, small, lightweight, low power device. The TAC-BIO detector's response time was one minute from when it sensed a biological agent. It was also demonstrated that the detector could be operated unattended indoors and outdoors for weeks at a time.

Aerosolized biological particles will fluoresce and scatter light under a UV light beam. Observed fluorescence is dependent on the applied wavelength and the biochemical fluorophores within the biological agent. UV induced fluorescence offers a rapid, accurate, efficient and logistically practical way for biological agent detection. This is because the use of UV fluorescence is reagent less, or a process that does not require an added chemical to produce a reaction, with no consumables, or produces no chemical byproducts.

Additionally, TAC-BIO can reliably discriminate between threat and non-threat aerosols. It was claimed to be sensitive enough to detect low concentrations, but not so sensitive that it would cause false positives. The particle counting algorithm used in the device converted raw data into information by counting the photon pulses per unit of time from the fluorescence and scattering detectors, and comparing the value to a set threshold.

The original TAC-BIO was introduced in 2010, while the second generation TAC-BIO GEN II, was designed in 2015 to be more cost efficient as plastic parts were used. It's small, light-weight design allows it to be mounted to vehicles, robots, and unmanned aerial vehicles. The second generation device could also be utilized as an environmental detector to monitor air quality in hospitals, airplanes, or even in households to detect fungus and mold.

Other Applications

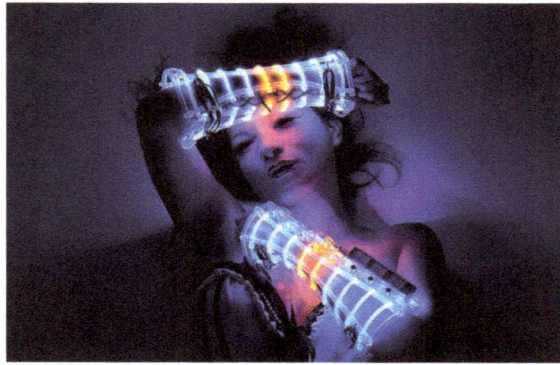

LED costume for stage performers.

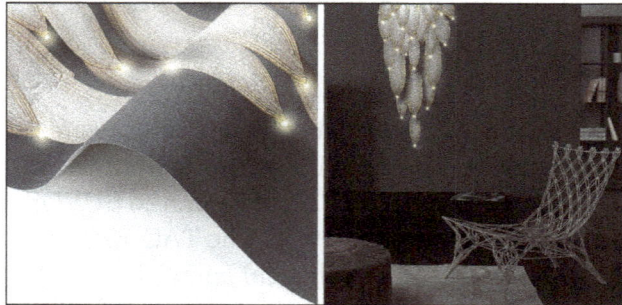

LED wallpaper by Meystyle.

The light from LEDs can be modulated very quickly so they are used extensively in optical fiber and free space optics communications. This includes remote controls, such as for television sets, where infrared LEDs are often used. Opto-isolators use an LED combined with a photodiode or phototransistor to provide a signal path with electrical isolation between two circuits. This is especially useful in medical equipment where the signals from a low-voltage sensor circuit (usually battery-powered) in contact with a living organism must be electrically isolated from any possible electrical failure in a recording or monitoring device operating at potentially dangerous voltages. An optoisolator also lets information be transferred between circuits that don't share a common ground potential.

Many sensor systems rely on light as the signal source. LEDs are often ideal as a light source due to the requirements of the sensors. The Nintendo Wii's sensor bar uses infrared LEDs. Pulse oximeters use them for measuring oxygen saturation. Some flatbed scanners use arrays of RGB LEDs rather than the typical cold-cathode fluorescent lamp as the light source. Having independent control of three illuminated colors allows the scanner to calibrate itself for more accurate color balance, and there is no need for warm-up. Further, its sensors only need be monochromatic, since at any one time the page being scanned is only lit by one color of light.

Since LEDs can also be used as photodiodes, they can be used for both photo emission and detection. This could be used, for example, in a touchscreen that registers reflected light from a finger or stylus. Many materials and biological systems are sensitive to, or dependent on, light. Grow lights use LEDs to increase photosynthesis in plants,and bacteria and viruses can be removed from water and other substances using UV LEDs for sterilization.

Deep UV LEDs, with a spectra range 247 nm to 386 nm, have other applications, such as water/air purification, surface disinfection, epoxy curing, free-space nonline-of-sight communication, high performance liquid chromatography, UV curing and printing, phototherapy, medical/ analytical instrumentation, and DNA absorption.

LEDs have also been used as a medium-quality voltage reference in electronic circuits. The forward voltage drop (about 1.7 V for a red LED or 1.2V for an infrared) can be used instead of a Zener diode in low-voltage regulators. Red LEDs have the flattest I/V curve above the knee. Nitride-based LEDs have a fairly steep I/V curve and are useless for this purpose. Although LED forward voltage is far more current-dependent than a Zener diode, Zener diodes with breakdown voltages below 3 V are not widely available.

The progressive miniaturization of low-voltage lighting technology, such as LEDs and OLEDs, suitable to incorporate into low-thickness materials has fostered experimentation in combining light sources and wall covering surfaces for interior walls in the form of LED wallpaper.

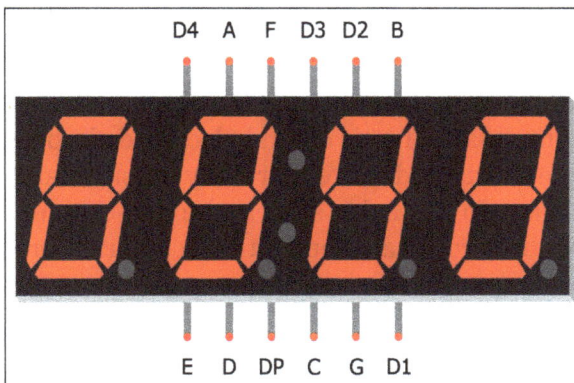

Seven-segment display that can display four digits and points.

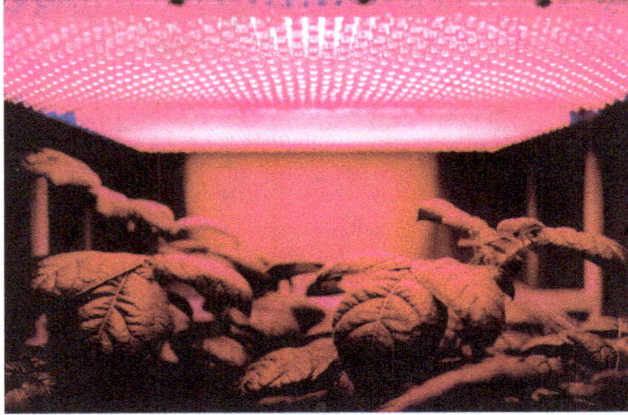

LED panel light source used in an experiment on plant growth. The findings of such experiments may be used to grow food in space on long duration missions.

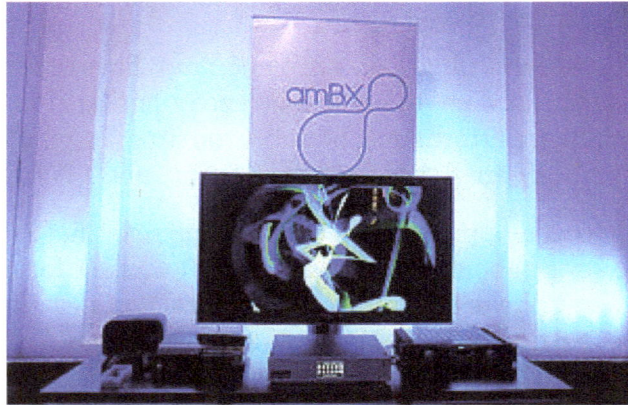

LED lights reacting dynamically to video feed via AmBX.

Research and Development

Key Challenges

LEDs require optimized efficiency to hinge on ongoing improvements such as phosphor materials and quantum dots.

The process of down-conversion (the method by which materials convert more-energetic photons to different, less energetic colors) also needs improvement. For example, the red phosphors that are used today are thermally sensitive and need to be improved in that aspect so that they do not color shift and experience efficiency drop-off with temperature. Red phosphors could also benefit from a narrower spectral width to emit more lumens and becoming more efficient at converting photons.

In addition, work remains to be done in the realms of current efficiency droop, color shift, system reliability, light distribution, dimming, thermal management, and power supply performance.

Potential Technology

Perovskite LEDs (PLEDs)

A new family of LEDs are based on the semiconductors called perovskites. In 2018, less than four years after their discovery, the ability of perovskite LEDs (PLEDs) to produce light from electrons already rivaled those of the best performing OLEDs. They have a potential for cost-effectiveness as they can be processed from solution, a low-cost and low-tech method, which might allow perovskite-based devices that have large areas to be made with extremely low cost. Their efficiency is superior by eliminating non-radiative losses, in other words, elimination of recombination pathways that do not produce photons; or by solving outcoupling problem (prevalent for thin-film LEDs) or balancing charge carrier injection to increase the EQE (external quantum efficiency). The most up-to-date PLED devices have broken the performance barrier by shooting the EQE above 20%.

In 2018, Cao and his colleagues' work, they targeted to solve the outcoupling problem, which is that the optical physics of thin-film LEDs causes the majority of light generated by the semiconductor to be trapped in the device. To achieve this goal, they demonstrated that solution-processed perovskites can spontaneously form submicrometre-scale crystal platelets, which can efficiently extract light from the device. These perovskites are formed simply by introducing amino-acif additives into the perovskite precursor solutions. In addition, their method is able to passivate perovskite surface defects and reduce nonradiative recombination. Therefore, by improving the light outcoupling and reducing nonradiative losses, Cao and his colleagues successfully achieved PLED with EQE up to 20.7%.

In Lin and his colleague's work, however, they used a different approach to generate high EQE. Instead of modifying the microstructure of perovskite layer, they chose to adopt a new strategy for managing the compositional distribution in the device—an approach that simultaneously provides high luminescence and balanced charge injection. In other words, they still used flat emissive layer, but tried to optimize the balance of electrons and holes injected into the perovskite, so as to make the most efficient use of the charge carriers. Moreover, in the perovskite layer, the crystals are perfectly enclosed by MABr additive (where MA is CH_3NH_3). The MABr shell passivates the nonradiative defects that would otherwise be present perovskite crystals, resulting in reduction of the nonradiative recombination. Therefore, by balancing charge injection and decreasing nonradiative losses, Lin and his colleagues developed PLED with EQE up to 20.3%.

Two-way LEDs

Devices called "nanorods" are a form of LEDs that can also detect and absorb light. They consist of a quantum dot directly contacting two semiconductor materials (instead of just one as in a traditional LED). One semiconductor allows movement of positive charge and one allows movement of negative charge. They can emit light, sense light, and collect energy. The nanorod gathers electrons while the quantum dot shell gathers positive charges so the dot emits light. When the voltage is switched the opposite process occurs and the dot absorbs light. By 2017 the only color developed was red.

QUANTUM DOT

Quantum dots (QDs) are tiny semiconductor particles a few nanometres in size, having optical and electronic properties that differ from larger particles due to quantum mechanics. They are a central topic in nanotechnology. When the quantum dots are illuminated by UV light, an electron in the quantum dot can be excited to a state of higher energy. In the case of a semiconducting quantum dot, this process corresponds to the transition of an electron from the valence band to the conductance band. The excited electron can drop back into the valence band releasing its energy by the emission of light. This light emission (photoluminescence) is illustrated in the figure on the right. The color of that light depends on the energy difference between the conductance band and the valence band.

In the language of materials science, nanoscale semiconductor materials tightly confine either electrons or electron holes. Quantum dots are sometimes referred to as artificial atoms, emphasizing their singularity, having bound, discrete electronic states, like naturally occurring atoms or molecules.

Quantum dots have properties intermediate between bulk semiconductors and discrete atoms or molecules. Their optoelectronic properties change as a function of both size and shape. Larger QDs of 5–6 nm diameter emit longer wavelengths, with colors such as orange or red. Smaller QDs (2–3 nm) emit shorter wavelengths, yielding colors like blue and green. However, the specific colors vary depending on the exact composition of the QD.

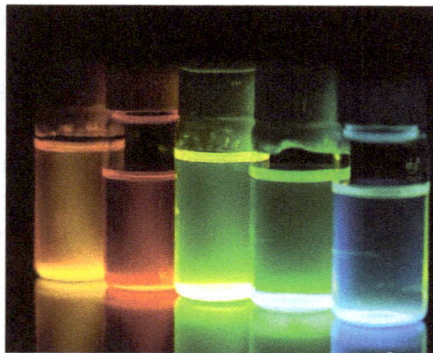

Colloidal quantum dots irradiated with a UV light. Different sized quantum dots emit different color light due to quantum confinement.

Potential applications of quantum dots include single-electron transistors, solar cells, LEDs, lasers, single-photon sources, second-harmonic generation, quantum computing,and medical imaging. Their small size allows for some QDs to be suspended in solution, which may lead to use in inkjet printing and spin-coating. They have been used in Langmuir-Blodgett thin-films. These processing techniques result in less expensive and less time-consuming methods of semiconductor fabrication.

Production

There are several ways to fabricate quantum dots. Possible methods include colloidal synthesis, self-assembly, and electrical gating.

Colloidal Synthesis

Colloidal semiconductor nanocrystals are synthesized from solutions, much like traditional chemical processes. The main difference is the product neither precipitates as a bulk solid nor remains dissolved. Heating the solution at high temperature, the precursors decompose forming monomers which then nucleate and generate nanocrystals. Temperature is a critical factor in determining optimal conditions for the nanocrystal growth. It must be high enough to allow for rearrangement and annealing of atoms during the synthesis process while being low enough to promote crystal growth. The concentration of monomers is another critical factor that has to be stringently controlled during nanocrystal growth. The growth process of nanocrystals can occur in two different regimes, "focusing" and "defocusing". At high monomer concentrations, the critical size (the size where nanocrystals neither grow nor shrink) is relatively small, resulting in growth of nearly all particles. In this regime, smaller particles grow faster than large ones (since larger crystals need more atoms to grow than small crystals) resulting in "focusing" of the size distribution, yielding an improbable distribution of nearly monodispersed particles. The size focusing is optimal when the monomer concentration is kept such that the average nanocrystal size present is always slightly larger than the critical size. Over time, the monomer concentration diminishes, the critical size becomes larger than the average size present, and the distribution "defocuses".

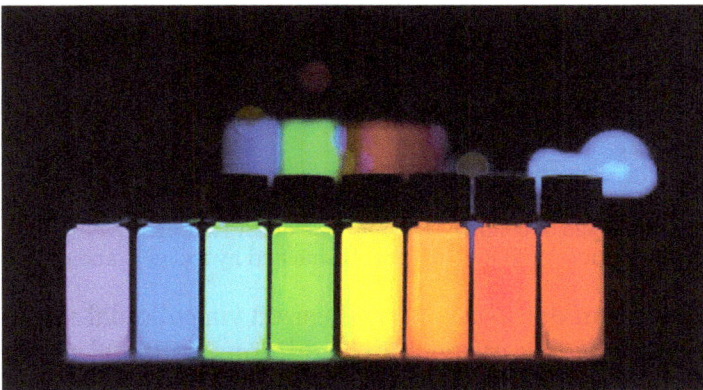

Quantum dots with gradually stepping emission from violet to deep red.

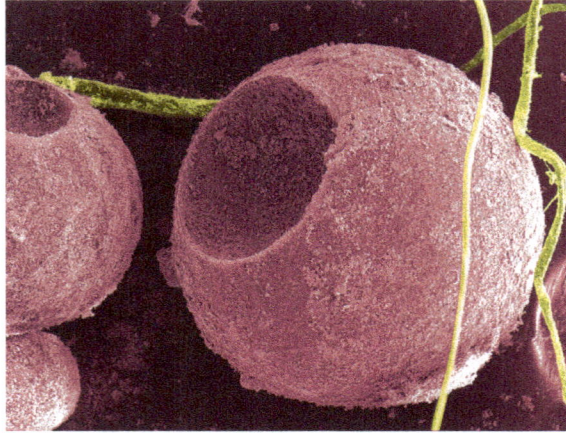

Cadmium sulfide quantum dots on cells.

There are colloidal methods to produce many different semiconductors. Typical dots are made of binary compounds such as lead sulfide, lead selenide, cadmium selenide, cadmium sulfide, cadmium telluride, indium arsenide, and indium phosphide. Dots may also be made from ternary compounds such as cadmium selenide sulfide. Further, recent advances have been made which allow for synthesis of colloidal perovskite quantum dots. These quantum dots can contain as few as 100 to 100,000 atoms within the quantum dot volume, with a diameter of ≈10 to 50 atoms. This corresponds to about 2 to 10 nanometers, and at 10 nm in diameter, nearly 3 million quantum dots could be lined up end to end and fit within the width of a human thumb.

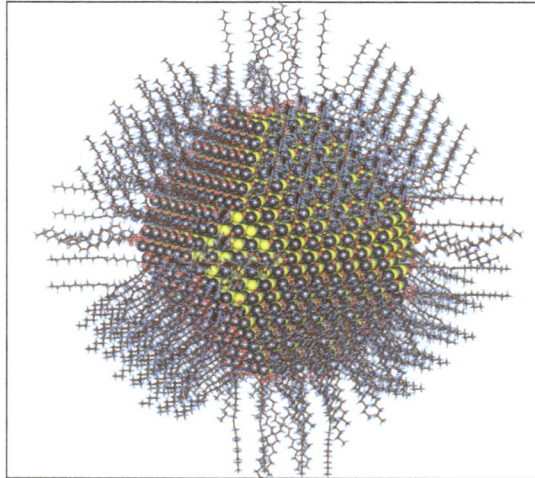

Idealized image of colloidal nanoparticle of lead sulfide (selenide) with
complete passivation by oleic acid, oleyl amine and hydroxyl ligands (size ≈5nm).

Large batches of quantum dots may be synthesized via colloidal synthesis. Due to this scalability and the convenience of benchtop conditions, colloidal synthetic methods are promising for commercial applications. It is acknowledged to be the least toxic of all the different forms of synthesis.

Plasma Synthesis

Plasma synthesis has evolved to be one of the most popular gas-phase approaches for the production of quantum dots, especially those with covalent bonds. For example, silicon (Si) and germanium (Ge) quantum dots have been synthesized by using non-thermal plasma. The size, shape, surface and composition of quantum dots can all be controlled in nonthermal plasma. Doping that seems quite challenging for quantum dots has also been realized in plasma synthesis. Quantum dots synthesized by plasma are usually in the form of powder, for which surface modification may be carried out. This can lead to excellent dispersion of quantum dots in either organic solventsor water(i. e. colloidal quantum dots).

Fabrication

- Self-assembled quantum dots are typically between 5 and 50 nm in size. Quantum dots defined by lithographically patterned gate electrodes, or by etching on two-dimensional electron gasses in semiconductor heterostructures can have lateral dimensions between 20 and 100 nm.

- Some quantum dots are small regions of one material buried in another with a larger band gap. These can be so-called core–shell structures, e.g. with CdSe in the core and ZnS in the shell, or from special forms of silica called ormosil. Sub-monolayer shells can also be effective ways of passivating the quantum dots, such as PbS cores with sub-monolayer CdS shells.

- Quantum dots sometimes occur spontaneously in quantum well structures due to monolayer fluctuations in the well's thickness.

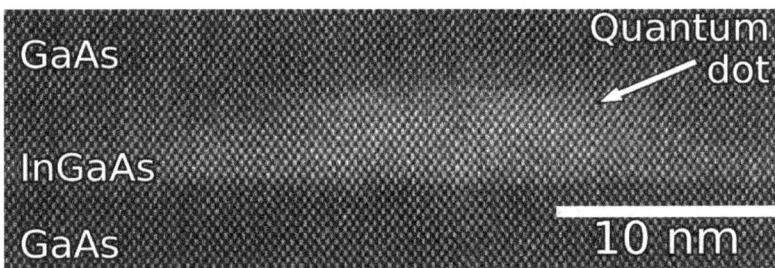

Atomic resolution scanning transmission electron microscopy
image of an InGaAs quantum dot buried in GaAs.

- Self-assembled quantum dots nucleate spontaneously under certain conditions during molecular beam epitaxy (MBE) and metalorganic vapour-phase epitaxy (MOVPE), when a material is grown on a substrate to which it is not lattice matched. The resulting strain leads to the formation of islands on top of a two-dimensional wetting layer. This growth mode is known as Stranski–Krastanov growth. The islands can be subsequently buried to form the quantum dot. A widely used type of quantum dots grown with this method are In(Ga)As quantum dots in GaAs.Such quantum dots have the potential for applications in quantum cryptography (i.e. single photon sources) and quantum computation. The main limitations of this method are the cost of fabrication and the lack of control over positioning of individual dots.

- Individual quantum dots can be created from two-dimensional electron or hole gases present in remotely doped quantum wells or semiconductor heterostructures called lateral quantum dots. The sample surface is coated with a thin layer of resist. A lateral pattern is then defined in the resist by electron beam lithography. This pattern can then be transferred to the electron or hole gas by etching, or by depositing metal electrodes (lift-off process) that allow the application of external voltages between the electron gas and the electrodes. Such quantum dots are mainly of interest for experiments and applications involving electron or hole transport, i.e. an electrical current.

- The energy spectrum of a quantum dot can be engineered by controlling the geometrical size, shape, and the strength of the confinement potential. Also, in contrast to atoms, it is relatively easy to connect quantum dots by tunnel barriers to conducting leads, which allows the application of the techniques of tunneling spectroscopy for their investigation.

The quantum dot absorption features correspond to transitions between discrete, three-dimensional particle in a box states of the electron and the hole, both confined to the same nanometer-size box. These discrete transitions are reminiscent of atomic spectra and have resulted in quantum dots also being called *artificial atoms*

- Confinement in quantum dots can also arise from electrostatic potentials (generated by external electrodes, doping, strain, or impurities).

- Complementary metal-oxide-semiconductor (CMOS) technology can be employed to fabricate silicon quantum dots. Ultra small (L=20 nm, W=20 nm) CMOS transistors behave as single electron quantum dots when operated at cryogenic temperature over a range of −269 °C (4 K) to about −258 °C (15 K). The transistor displays Coulomb blockade due to progressive charging of electrons one by one. The number of electrons confined in the channel is driven by the gate voltage, starting from an occupation of zero electrons, and it can be set to 1 or many.

Viral Assembly

Genetically engineered M13 bacteriophage viruses allow preparation of quantum dot biocomposite structures. It had previously been shown that genetically engineered viruses can recognize specific semiconductor surfaces through the method of selection by combinatorial phage display. Additionally, it is known that liquid crystalline structures of wild-type viruses (Fd, M13, and TMV) are adjustable by controlling the solution concentrations, solution ionic strength, and the external magnetic field applied to the solutions. Consequently, the specific recognition properties of the virus can be used to organize inorganic nanocrystals, forming ordered arrays over the length scale defined by liquid crystal formation. Using this information, Lee et al. were able to create self-assembled, highly oriented, self-supporting films from a phage and ZnS precursor solution. This system allowed them to vary both the length of bacteriophage and the type of inorganic material through genetic modification and selection.

Electrochemical Assembly

Highly ordered arrays of quantum dots may also be self-assembled by electrochemical techniques. A template is created by causing an ionic reaction at an electrolyte-metal interface which results in the spontaneous assembly of nanostructures, including quantum dots, onto the metal which is then used as a mask for mesa-etching these nanostructures on a chosen substrate.

Bulk-manufacture

Quantum dot manufacturing relies on a process called "high temperature dual injection" which has been scaled by multiple companies for commercial applications that require large quantities (hundreds of kilograms to tonnes) of quantum dots. This reproducible production method can be applied to a wide range of quantum dot sizes and compositions.

The bonding in certain cadmium-free quantum dots, such as III-V-based quantum dots, is more covalent than that in II-VI materials, therefore it is more difficult to separate nanoparticle nucleation and growth via a high temperature dual injection synthesis. An alternative method of quantum dot synthesis, the "molecular seeding" process, provides a reproducible route to the production of high quality quantum dots in large volumes. The process utilises identical molecules of a molecular cluster compound as the nucleation sites for nanoparticle growth, thus avoiding the need for a high temperature injection step. Particle growth is maintained by the periodic addition of precursors at moderate temperatures until the desired particle size is reached. The molecular seeding process is not limited to the production of cadmium-free quantum dots; for example, the process can be used to synthesise kilogram batches of high quality II-VI quantum dots in just a few hours.

Another approach for the mass production of colloidal quantum dots can be seen in the transfer of the well-known hot-injection methodology for the synthesis to a technical continuous flow system. The batch-to-batch variations arising from the needs during the mentioned methodology can be overcome by utilizing technical components for mixing and growth as well as transport and temperature adjustments. For the production of CdSe based semiconductor nanoparticles this method has been investigated and tuned to production amounts of kg per month. Since the use of technical components allows for easy interchange in regards of maximum through-put and size, it can be further enhanced to tens or even hundreds of kilograms.

Heavy-metal-free Quantum Dots

In many regions of the world there is now a restriction or ban on the use of heavy metals in many household goods, which means that most cadmium-based quantum dots are unusable for consumer-goods applications.

For commercial viability, a range of restricted, heavy-metal-free quantum dots has been developed showing bright emissions in the visible and near infra-red region of the spectrum and have similar optical properties to those of CdSe quantum dots. Among these systems are InP/ZnS and CuInS/ZnS, for example.

Peptides are being researched as potential quantum dot material.Since peptides occur naturally in all organisms, such dots would likely be nontoxic and easily biodegraded.

Optical Properties

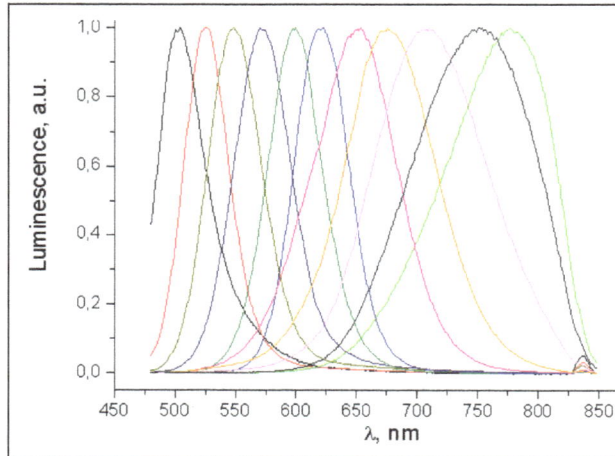

Fluorescence spectra of CdTe quantum dots of various sizes. Different sized
quantum dots emit different color light due to quantum confinement.

In semiconductors, light absorption generally leads to an electron being excited from the valence to the conduction band, leaving behind a hole. The electron and the hole can bind to each other to form an exciton. When this exciton recombines (i.e. the electron resumes its ground state), the exciton's energy can be emitted as light. This is called fluorescence. In a simplified model, the energy of the emitted photon can be understood as the sum of the band gap energy between the highest occupied level and the lowest unoccupied energy level, the confinement energies of the hole and the excited electron, and the bound energy of the exciton (the electron-hole pair).

As the confinement energy depends on the quantum dot's size, both absorption onset and fluorescence emission can be tuned by changing the size of the quantum dot during its synthesis. The larger the dot, the redder (lower energy) its absorption onset and fluorescence spectrum. Conversely, smaller dots absorb and emit bluer (higher energy) light.

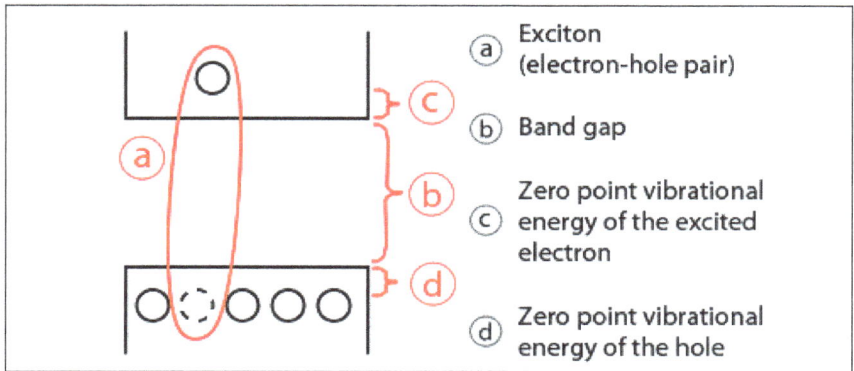

To improve fluorescence quantum yield, quantum dots can be made with "shells" of a larger bandgap semiconductor material around them. The improvement is suggested to be due to the reduced access of electron and hole to non-radiative surface recombination pathways in some cases, but also due to reduced Auger recombination in others.

Potential Applications

Quantum dots are particularly promising for optical applications due to their high extinction coefficient. They operate like a single electron transistor and show the Coulomb blockade effect. Quantum dots have also been suggested as implementations of qubits for quantum information processing,and as active elements for thermoelectrics.

Tuning the size of quantum dots is attractive for many potential applications. For instance, larger quantum dots have a greater spectrum-shift towards red compared to smaller dots, and exhibit less pronounced quantum properties. Conversely, the smaller particles allow one to take advantage of more subtle quantum effects.

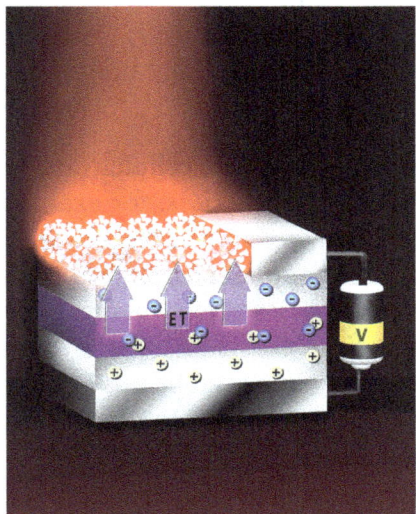

A device that produces visible light, through energy transfer from thin layers of quantum wells to crystals above the layers.

Being zero-dimensional, quantum dots have a sharper density of states than higher-dimensional structures. As a result, they have superior transport and optical properties. They have potential uses in diode lasers, amplifiers, and biological sensors. Quantum dots may be excited within a locally enhanced electromagnetic field produced by gold nanoparticles, which can then be observed from the surface plasmon resonance in the photoluminescent excitation spectrum of (CdSe)ZnS nanocrystals. High-quality quantum dots are well suited for optical encoding and multiplexing applications due to their broad excitation profiles and narrow/symmetric emission spectra. The new generations of quantum dots have far-reaching potential for the study of intracellular processes at the single-molecule level, high-resolution cellular imaging, long-term in vivo observation of cell trafficking, tumor targeting, and diagnostics.

CdSe nanocrystals are efficient triplet photosensitizers. Laser excitation of small CdSe nanoparticles enables the extraction of the excited state energy from the Quantum Dots into bulk solution, thus opening the door to a wide range of potential applications such as photodynamic therapy, photovoltaic devices, molecular electronics, and catalysis.

Biology

In modern biological analysis, various kinds of organic dyes are used. However, as technology advances, greater flexibility in these dyes is sought. To this end, quantum dots have quickly filled in the role, being found to be superior to traditional organic dyes on several counts, one of the most immediately obvious being brightness (owing to the high extinction coefficient combined with a comparable quantum yield to fluorescent dyes) as well as their stability (allowing much less photobleaching). It has been estimated that quantum dots are 20 times brighter and 100 times more stable than traditional fluorescent reporters. For single-particle tracking, the irregular blinking of quantum dots is a minor drawback. However, there have been groups which have developed quantum dots which are essentially nonblinking and demonstrated their utility in single molecule tracking experiments.

The use of quantum dots for highly sensitive cellular imaging has seen major advances. The improved photostability of quantum dots, for example, allows the acquisition of many consecutive focal-plane images that can be reconstructed into a high-resolution three-dimensional image. Another application that takes advantage of the extraordinary photostability of quantum dot probes is the real-time tracking of molecules and cells over extended periods of time. Antibodies, streptavidin, peptides, DNA, nucleic acid aptamers, or small-molecule ligands can be used to target quantum dots to specific proteins on cells. Researchers were able to observe quantum dots in lymph nodes of mice for more than 4 months.

Quantum dots can have antibacterial properties similar to nanoparticles and can kill bacteria in a dose-dependent manner. One mechanism by which quantum dots can kill bacteria is through impairing the functions of antioxidative system in the cells and down regulating the antioxidative genes. In addition, quantum dots can directly damage the cell wall. Quantum dots have been shown to be effective against both gram-positive and gram-negative bacteria.

Semiconductor quantum dots have also been employed for in vitro imaging of pre-labeled cells. The ability to image single-cell migration in real time is expected to be important to several research areas such as embryogenesis, cancer metastasis, stem cell therapeutics, and lymphocyte immunology.

One application of quantum dots in biology is as donor fluorophores in Förster resonance energy transfer, where the large extinction coefficient and spectral purity of these fluorophores make them superior to molecular fluorophores. It is also worth noting that the broad absorbance of QDs allows selective excitation of the QD donor and a minimum excitation of a dye acceptor in FRET-based studies. The applicability of the FRET model, which assumes that the Quantum Dot can be approximated as a point dipole, has recently been demonstrated.

The use of quantum dots for tumor targeting under in vivo conditions employ two targeting schemes: active targeting and passive targeting. In the case of active targeting, quantum dots are functionalized with tumor-specific binding sites to selectively bind to tumor cells. Passive targeting uses the enhanced permeation and retention of tumor cells for the delivery of quantum dot probes. Fast-growing tumor cells typically have more permeable membranes than healthy cells, allowing the leakage of small nanoparticles into the cell body. Moreover, tumor cells lack an effective lymphatic drainage system, which leads to subsequent nanoparticle-accumulation.

Quantum dot probes exhibit in vivo toxicity. For example, CdSe nanocrystals are highly toxic to cultured cells under UV illumination, because the particles dissolve, in a process known as photolysis, to release toxic cadmium ions into the culture medium. In the absence of UV irradiation, however, quantum dots with a stable polymer coating have been found to be essentially nontoxic. Hydrogel encapsulation of quantum dots allows for quantum dots to be introduced into a stable aqueous solution, reducing the possibility of cadmium leakage. Then again, only little is known about the excretion process of quantum dots from living organisms.

In another potential application, quantum dots are being investigated as the inorganic fluorophore for intra-operative detection of tumors using fluorescence spectroscopy.

Delivery of undamaged quantum dots to the cell cytoplasm has been a challenge with existing techniques. Vector-based methods have resulted in aggregation and endosomal sequestration of quantum dots while electroporation can damage the semi-conducting particles and aggregate delivered dots in the cytosol. Via cell squeezing, quantum dots can be efficiently delivered without inducing aggregation, trapping material in endosomes, or significant loss of cell viability. Moreover, it has shown that individual quantum dots delivered by this approach are detectable in the cell cytosol, thus illustrating the potential of this technique for single molecule tracking studies.

Photovoltaic Devices

The tunable absorption spectrum and high extinction coefficients of quantum dots make them attractive for light harvesting technologies such as photovoltaics. Quantum dots may be able to increase the efficiency and reduce the cost of today's typical silicon photovoltaic cells. According to an experimental proof from 2004, quantum dots of lead selenide can produce more than one exciton from one high energy photon via the process of carrier multiplication or multiple exciton generation (MEG). This compares favorably to today's photovoltaic cells which can only manage one exciton per high-energy photon, with high kinetic energy carriers losing their energy as heat. Quantum dot photovoltaics would theoretically be cheaper to manufacture, as they can be made "using simple chemical reactions."

Quantum Dot only Solar Cells

Aromatic self-assembled monolayers (SAMs) (e.g. 4-nitrobenzoic acid) can be used to improve the band alignment at electrodes for better efficiencies. This technique has provided a record power conversion efficiency (PCE) of 10.7%. The SAM is positioned between ZnO-PbS colloidal quantum dot (CQD) film junction to modify band alignment via the dipole moment of the constituent SAM molecule, and the band tuning may be modified via the density, dipole and the orientation of the SAM molecule.

Quantum Dot in Hybrid Solar Cells

Colloidal quantum dots are also used in inorganic/organic hybrid solar cells. These solar cells are attractive because of the potential for low-cost fabrication and relatively high efficiency. Incorporation of metal oxides, such as ZnO, TiO_2, and Nb_2O_5 nanomaterials into organic photovoltaics have been commercialized using full roll-to-roll processing. A 13.2% power conversion efficiency is claimed in Si nanowire/PEDOT:PSS hybrid solar cells.

Quantum Dot with Nanowire in Solar Cells

Another potential use involves capped single-crystal ZnO nanowires with CdSe quantum dots, immersed in mercaptopropionic acid as hole transport medium in order to obtain a QD-sensitized solar cell. The morphology of the nanowires allowed the electrons to have a direct pathway to the photoanode. This form of solar cell exhibits 50–60% internal quantum efficiencies.

Nanowires with quantum dot coatings on silicon nanowires (SiNW) and carbon quantum dots. The use of SiNWs instead of planar silicon enhances the antiflection properties of Si.The SiNW exhibits a light-trapping effect due to light trapping in the SiNW. This use of SiNWs in conjunction with carbon quantum dots resulted in a solar cell that reached 9.10% PCE.

Graphene quantum dots have also been blended with organic electronic materials to improve efficiency and lower cost in photovoltaic devices and organic light emitting diodes (OLEDs) in compared to graphene sheets. These graphene quantum dots were functionalized with organic ligands that experience photoluminescence from UV-Vis absorption.

Light Emitting Diodes

Several methods are proposed for using quantum dots to improve existing light-emitting diode (LED) design, including "Quantum Dot Light Emitting Diode" (QD-LED or QLED) displays and "Quantum Dot White Light Emitting Diode" (QD-WLED) displays. Because Quantum dots naturally produce monochromatic light, they can be more efficient than light sources which must be color filtered. QD-LEDs can be fabricated on a silicon substrate, which allows them to be integrated onto standard silicon-based integrated circuits or microelectromechanical systems.

Quantum Dot Displays

Quantum dots are valued for displays because they emit light in very specific gaussian distributions. This can result in a display with visibly more accurate colors.

A conventional color liquid crystal display (LCD) is usually backlit by fluorescent lamps (CCFLs) or conventional white LEDs that are color filtered to produce red, green, and blue pixels. Quantum dot displays use blue-emitting LEDs rather than white LEDs as the light sources. The converting part of the emitted light is converted into pure green and red light by the corresponding color quantum dots placed in front of the blue LED or using a quantum dot infused diffuser sheet in the backlight optical stack. Blank pixels are also used to allow the blue LED light to still generate blue hues. This type of white light as the backlight of an LCD panel allows for the best color gamut at lower cost than an RGB LED combination using three LEDs.

Another method by which quantum dot displays can be achieved is the electroluminescent (EL) or electro-emissive method. This involves embedding quantum dots in each individual pixel. These are then activated and controlled via an electric current application.Since this is often light emitting itself, the achievable colors may be limited in this method.Electro-emissive QD-LED TVs exist in laboratories only.

The ability of QDs to precisely convert and tune a spectrum makes them attractive for LCD displays. Previous LCD displays can waste energy converting red-green poor, blue-yellow rich white light into a more balanced lighting. By using QDs, only the necessary colors for ideal images are contained in the screen. The result is a screen that is brighter, clearer, and more energy-efficient. The first commercial application of quantum dots was the Sony XBR X900A series of flat panel televisions released in 2013.

In June 2006, QD Vision announced technical success in making a proof-of-concept quantum dot display and show a bright emission in the visible and near infra-red region of the spectrum. A QD-LED integrated at a scanning microscopy tip was used to demonstrate fluorescence near-field scanning optical microscopy (NSOM) imaging.

Photodetector Devices

Quantum dot photodetectors (QDPs) can be fabricated either via solution-processing,or from conventional single-crystalline semiconductors.Conventional single-crystalline semiconductor QDPs are precluded from integration with flexible organic electronics due to the incompatibility of their growth conditions with the process windows required by organic semiconductors. On the other hand, solution-processed QDPs can be readily integrated with an almost infinite variety of substrates, and also postprocessed atop other integrated circuits. Such colloidal QDPs have potential applications in surveillance, machine vision, industrial inspection, spectroscopy, and fluorescent biomedical imaging.

Photocatalysts

Quantum dots also function as photocatalysts for the light driven chemical conversion of water into hydrogen as a pathway to solar fuel. In photocatalysis, electron hole pairs formed in the dot under band gap excitation drive redox reactions in the surrounding liquid. Generally, the photocatalytic activity of the dots is related to the particle size and its degree of quantum confinement. This is because the band gap determines the chemical energy that is stored in the dot in the excited state. An obstacle for the use of quantum dots in photocatalysis is the presence of surfactants on the surface of the dots. These surfactants (or ligands) interfere with the chemical reactivity of the dots by slowing down mass transfer and electron transfer processes. Also, quantum dots made of metal chalcogenides are chemically unstable under oxidizing conditions and undergo photo corrosion reactions.

Theory

Quantum dots are theoretically described as a point like, or a zero dimensional (0D) entity. Most of their properties depend on the dimensions, shape and materials of which QDs are made. Generally QDs present different thermodynamic properties from the bulk materials of which they are made. One of these effects is the

Melting-point depression. Optical properties of spherical metallic QDs are well described by the Mie scattering theory.

Quantum Confinement in Semiconductors

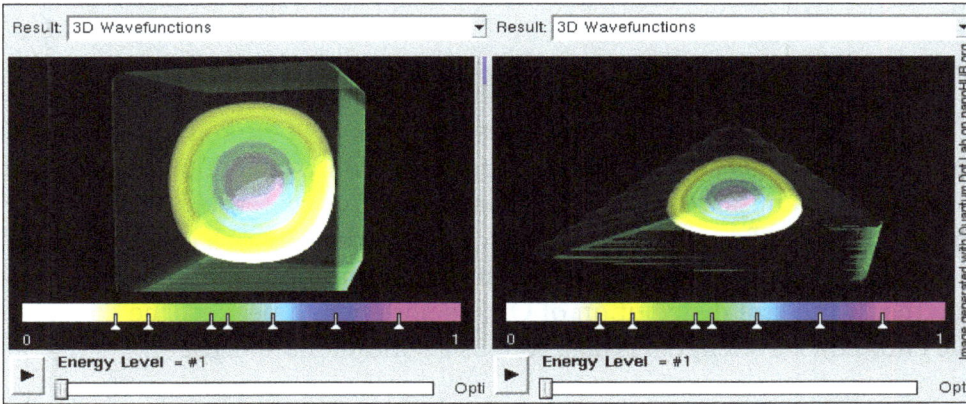

3D confined electron wave functions in a quantum dot. Here, rectangular and triangular-shaped quantum dots are shown. Energy states in rectangular dots are more *s-type* and *p-type*. However, in a triangular dot the wave functions are mixed due to confinement symmetry.

In a semiconductor crystallite whose size is smaller than twice the size of its exciton Bohr radius, the excitons are squeezed, leading to quantum confinement. The energy levels can then be predicted using the particle in a box model in which the energies of states depend on the length of the box. Comparing the quantum dots size to the Bohr radius of the electron and hole wave functions, 3 regimes can be defined. A 'strong confinement regime' is defined as the quantum dots radius being smaller than both electron and hole Bohr radius, 'weak confinement' is given when the quantum dot is larger than both. For semiconductors in which electron and hole radii are markedly different, an 'intermediate confinement regime' exists, where the quantum dot's radius is larger than the Bohr radius of one charge carrier (typically the hole), but not the other charge carrier.

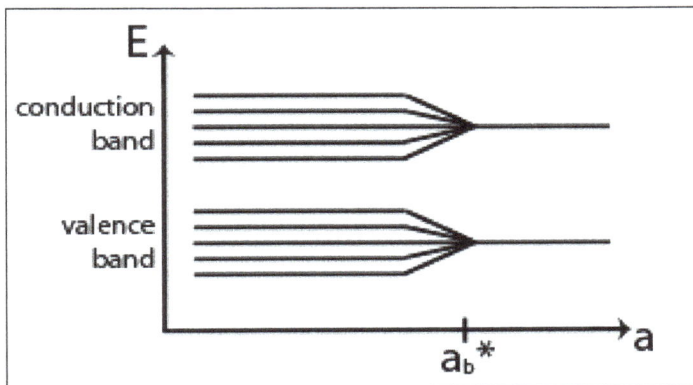

Splitting of energy levels for small quantum dots due to the quantum confinement effect. The horizontal axis is the radius, or the size, of the quantum dots and $a_b{}^*$ is the Exciton Bohr radius.

Band Gap Energy

The band gap can become smaller in the strong confinement regime as the energy levels split up. The Exciton Bohr radius can be expressed as:

$$a_b^* = \varepsilon_r \left(\frac{m}{\mu} \right) a_b$$

where a_b is the Bohr radius=0.053 nm, m is the mass, μ is the reduced mass, and ε_r is the size-dependent dielectric constant (Relative permittivity). This results in the increase in the total emission energy (the sum of the energy levels in the smaller band gaps in the strong confinement regime is larger than the energy levels in the band gaps of the original levels in the weak confinement regime) and the emission at various wavelengths. If the size distribution of QDs is not enough peaked, the convolution of multiple emission wavelengths is observed as a continuous spectra.

Confinement Energy

The exciton entity can be modeled using the particle in the box. The electron and the hole can be seen as hydrogen in the Bohr model with the hydrogen nucleus replaced by the hole of positive charge and negative electron mass. Then the energy levels of the exciton can be represented as the solution to the particle in a box at the ground level (n = 1) with the mass replaced by the reduced mass. Thus by varying the size of the quantum dot, the confinement energy of the exciton can be controlled.

Bound Exciton Energy

There is Coulomb attraction between the negatively charged electron and the positively charged hole. The negative energy involved in the attraction is proportional to Rydberg's energy and inversely proportional to square of the size-dependent dielectric constant of the semiconductor. When the size of the semiconductor crystal is smaller than the Exciton Bohr radius, the Coulomb interaction must be modified to fit the situation.

Therefore, the sum of these energies can be represented as:

$$E_{confinement} = \frac{\hbar^2 \pi^2}{2a^2} \left(\frac{1}{m_e} + \frac{1}{m_h} \right) = \frac{\hbar^2 \pi^2}{2 \mu a^2}$$

$$E_{exciton} = -\frac{1}{\epsilon_r^2} \frac{\mu}{m_e} R_y = -R_y^*$$

$$E = E_{bandgap} + E_{confinement} + E_{exciton}$$

$$= E_{bandgap} + \frac{\hbar^2 \pi^2}{2 \mu a^2} - R_y^*$$

where μ is the reduced mass, a is the radius of the quantum dot, m_e is the free electron mass, m_h is the hole mass, and ε_r is the size-dependent dielectric constant.

Although the above equations were derived using simplifying assumptions, they imply that the electronic transitions of the quantum dots will depend on their size. These quantum confinement effects are apparent only below the critical size. Larger particles do not exhibit this effect. This effect of quantum confinement on the quantum dots has been repeatedly verified experimentallyand is a key feature of many emerging electronic structures.

The Coulomb interaction between confined carriers can also be studied by numerical means when results unconstrained by asymptotic approximations are pursued.

Besides confinement in all three dimensions (i.e., a quantum dot), other quantum confined semiconductors include:

- Quantum wires, which confine electrons or holes in two spatial dimensions and allow free propagation in the third.

- Quantum wells, which confine electrons or holes in one dimension and allow free propagation in two dimensions.

Models

A variety of theoretical frameworks exist to model optical, electronic, and structural properties of quantum dots. These may be broadly divided into quantum mechanical, semiclassical, and classical.

Quantum Mechanics

Quantum mechanical models and simulations of quantum dots often involve the interaction of electrons with a pseudopotential or random matrix.

Semiclassical

Semiclassical models of quantum dots frequently incorporate a chemical potential. For example, the thermodynamic chemical potential of an N-particle system is given by:

$$\mu(N) = E(N) - E(N-1)$$

whose energy terms may be obtained as solutions of the Schrödinger equation. The definition of capacitance,

$$\frac{1}{C} = \frac{\Delta V}{\Delta Q},$$

with the potential difference:

$$\Delta V = \frac{\Delta \mu}{e} = \frac{\mu(N + \Delta N) - \mu(N)}{e}$$

may be applied to a quantum dot with the addition or removal of individual electrons,

$$\Delta N = 1 \text{ and } \Delta Q = e$$

Then,

$$C(N) = \frac{e^2}{\mu(N+1) - \mu(N)} = \frac{e^2}{I(N) - A(N)}$$

is the "quantum capacitance" of a quantum dot, where we denoted by $I(N)$ the ionization potential and by $A(N)$ the electron affinity of the N-particle system.

Classical Mechanics

Classical models of electrostatic properties of electrons in quantum dots are similar in nature to the Thomson problem of optimally distributing electrons on a unit sphere.

The classical electrostatic treatment of electrons confined to spherical quantum dots is similar to their treatment in the Thomson,or plum pudding model, of the atom.

The classical treatment of both two-dimensional and three-dimensional quantum dots exhibit electron shell-filling behavior. A "periodic table of classical artificial atoms" has been described for two-dimensional quantum dots. As well, several connections have been reported between the three-dimensional Thomson problem and electron shell-filling patterns found in naturally-occurring atoms found throughout the periodic table. This latter work originated in classical electrostatic modeling of electrons in a spherical quantum dot represented by an ideal dielectric sphere.

SILICON DRIFT DETECTOR

Silicon Drift Detectors (SDDs) are the current stateof-the-art for high resolution, high count rate X-ray spectroscopy. Modern SDDs benefit from a unique design that enables them to achieve a much higher performance than lithium drifted silicon – or Si(Li) – detectors. Specifically, they experience far less electronic noise, which is particularly observed at short peaking times (i.e. high count rates), larger detector areas and low X-ray energies. As a result, they have largely displaced Si(Li) detectors and are now used in large quantities for industrial scale applications like electron microscopy (SEM/EDS) and X-ray fluorescence analysis (EDXRF).

The figure above shows a typical SDD device. The chip is mounted on a thermoelectric cooler which provides device cooling. SDD devices come in a variety of shapes. The droplet-like shape shown in this figure is commonly used for smaller active areas such as 10 mm2 and 20 mm2. The radiation entrance window which is shown in the picture consists of a flat p-implanted region covered by a thin conductive layer to keep the entrance window radiation hard.

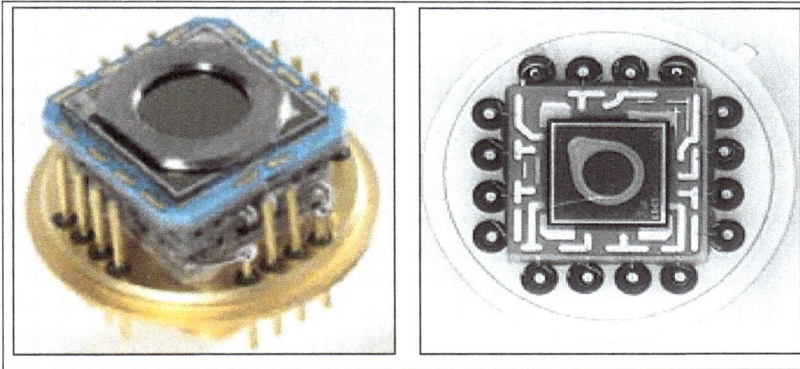

Typical mounted and bonded SDD devices.

X-ray Detector Fundamentals

Typical X-ray detection devices consist of an active region composed of fully depleted, high-resistivity silicon, a front contact area and a collection anode. X-rays incident upon the front contact area are absorbed in the bulk Si region and generate electron-hole pairs. The quantity of charged carriers generated depends on the energy of the incident X-ray. A pre-established electric field between the front contact and the anode causes these electrons and holes to drift along the field lines; i.e. toward the anode. The charge accumulated at the anode is then converted to a voltage by a pre-amplifier.

The incident X-ray energy can be determined by monitoring the magnitude of the voltage step after each pulse; i.e. after each incident X-ray is absorbed. Figure is an example schematic of the electronics associated with an X-ray detector. When the output waveform exhibits fluctuations due to noise, there is a limit to how precisely this voltage step is measured. The measurement imprecision creates a Gaussian spread for a given energy. Thus, noise effectively widens the measured X-ray peaks. Figure illustrates the voltage steps and associated noise. Noise is influenced by factors such as the FET gain, input capacitance and pre-amplifier leakage current. Over longer shaping times, the noise is averaged out and the resolution improves. Over shorter shaping times (i.e. those used to drive higher count rates), there is less averaging of the noise and subsequently more uncertainty in the voltage step, worsening resolution. Additionally, the signal to noise ratio is necessarily lower for low energy X-rays; indicating that noise plays a larger role in the resolution of low energy X-rays.

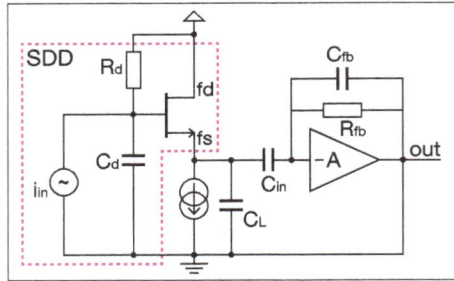

Schematic of the electronics in an X-ray detector. The SDD dashed line illustrates the electrical impact of the integrated FET.

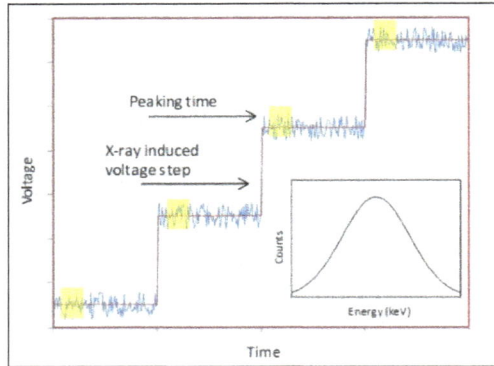

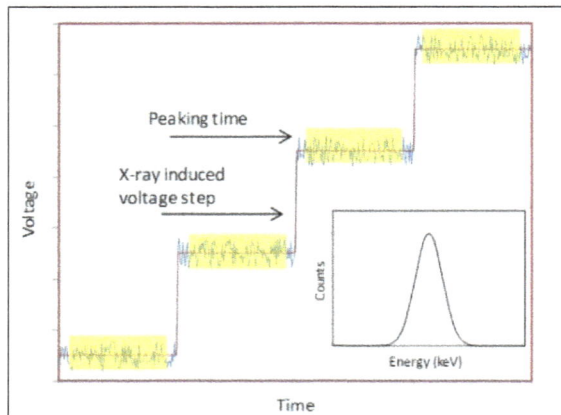

Illustration of voltage steps as a function of an absorbed X-ray. The noise fluctuations demonstrate the impact of noise and shaping time (top vs. bottom) on resolution.

Sources of Noise

There are several sources of electronic noise, characterized by the equations below.

Electronic noise $\propto$ shot noise + 1/f noise + thermal noise

$$\text{shot noise} \propto I_{\text{leak}}$$

$$1/f \text{ noise} \propto C_{\text{in}}^{2}$$

$$\text{thermal noise} \propto \frac{kT}{g_{\text{m}}} \frac{C_{\text{in}}^{2}}{\tau_{\text{peak}}}$$

The first factor is shot noise, generated by leakage current in the pre-amplifier. The next factor is "1/f" noise, which is directly related to the capacitance squared. The third factor is thermal noise, which is related to the capacitance squared, to the temperature and to the inverse peaking time. As the capacitance gets small enough – for example as the anode size is reduced – the total resolution for an X-ray becomes predominantly dependent on the shot noise. Because the shot noise is independent of temperature, cooling the detector is much less important to achieving good resolution. It is also now far less dependent on a long shaping time. Eventually the noise becomes small enough that resolution becomes almost entirely limited by Fano broadening. Fano broadening is based on statistical fluctuations in the radiation interaction with the Si crystal lattice and the charge production process. When this limit is reached, the theoretical best resolution is roughly 120 eV.

The figure demonstrates the resolution – as measured at the Mn Kα peak – as a function of shaping time for a typical SDD and a typical diode detector. Because the SDD has lower capacitance – and therefore lower noise – this translates into superior resolution with shorter shaping times and larger active areas; i.e. superior resolution at superior count rates.

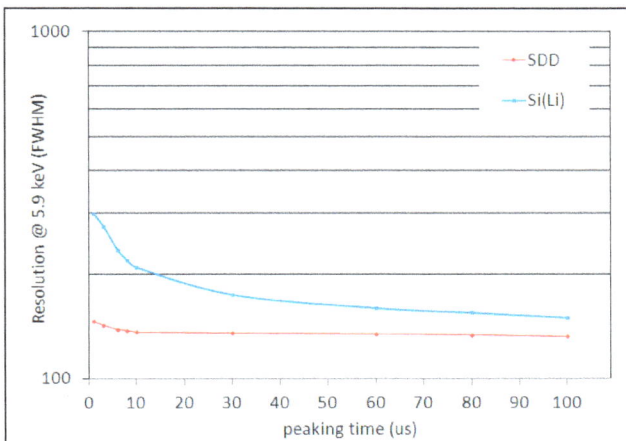

Energy resolution at Mn Kα as a function of peaking time for typical SDD and Si(Li) detectors.

Modern X-ray Detectors – the Silicon Drift Detector (SDD)

The basic form of the Silicon Drift Detector (SDD) was proposed in 1983 by Gatti & Rehak. It consists of a radial electric field which is intentionally established and controlled by a number of increasingly reverse-biased, circular field strips covering one surface of the device. This field terminates in a very small collecting anode on one face of the device. The design of an example SDD, figure, demonstrates this ring electrode structure, which creates the radial electric field. The radiation entrance window on the opposite side is composed of a thin, shallow implanted p+ doped region, which provides a homogeneous sensitivity over the whole detector area. A thin conductive layer is applied on top of this p+ doped region in order to improve radiation hardness.

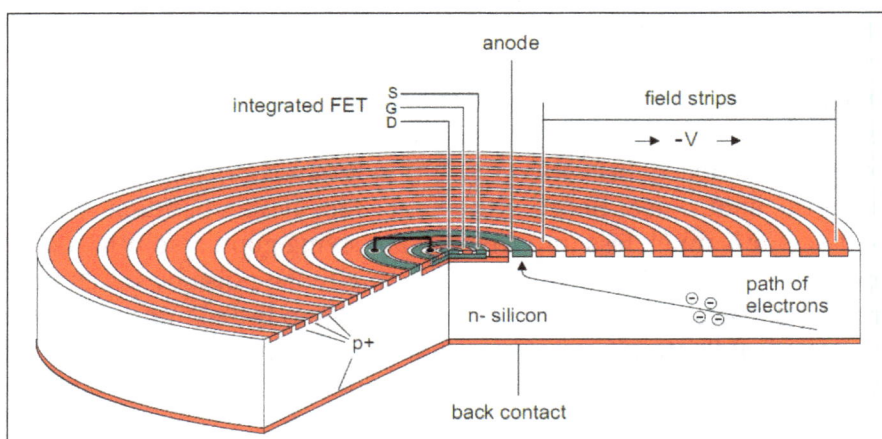

Example SDD detector design.

The unique value of this type of detector is the extremely small size of the anode, relative to the overall active area of the detector. The X-ray generated charged carriers (i.e. electrons and holes) are guided along these electric field lines to the very small anode at the center of the detector. Because the device capacitance is directly related to the size of the anode, a dramatically smaller anode results in a dramatically lower device capacitance. Typically observed anode capacitances are 25 – 150 fF. Because the electronic noise at short shaping times is proportional to capacitance squared, the benefit of a smaller anode is better resolution at shorter shaping times (higher count rates); in particular at low energies where the signal to noise is lower. The noise is small enough that the device can be operated at temperatures (~ -20 °C) that are readily achieved with a peltier device; thereby eliminating the need for LN_2 cooling.

To take full advantage of the small output capacitance, the front-end transistor of the amplifying electronics is integrated directly onto the detector chip and connected to the collecting anode by a short metal strip. This eliminates parasitic capacitance at bonding pads; minimizing capacitance between the detector anode and the amplifier FET. Additionally, noise caused by electric pickup, cross-talk and micro-phony effects are rendered insignificant. This impact is illustrated schematically in the figure.

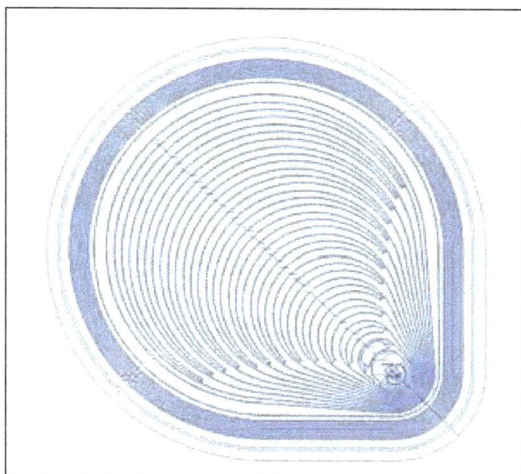

Modern detector with offset anode and pre-amplifier FET.

A more modern SDD design, figure, involves an offset anode and FET. This is often referred to as a "tear-drop" or "droplet" SDD. When the integrated FET is at the center of the device, as in figure, it is susceptible to irradiation by incident X-rays. Additionally, the electrostatic fields surrounding the FET result in performance losses at low X-ray energies. When the FET is offset, as in figure, it is outside the active area and therefore not subject to incoming radiation. This design is still relatively unique and is most common on smaller area (10 mm²) detectors.

References

- Lin, Che-I; Lai, Cheng-Hsiao; King, Ya-Chin (5 August 2004). "A four transistor CMOS active pixel sensor with high dynamic range operation". Proceedings of 2004 IEEE Asia-Pacific Conference on Advanced System Integrated Circuits: 124–127. doi:10.1109/APASIC.2004.1349425. ISBN 0-7803-8637-X

- Wilson, Matthew David; Cernik, Robert; Chen, Henry; Hansson, Conny; Iniewski, Kris; Jones, Lawrence L.; Seller, Paul; Veale, Matthew C. (2011). "Small pixel CZT detector for hard X-ray spectroscopy". Nuclear Instruments and Methods in Physics Research Section A: Accelerators, Spectrometers, Detectors and Associated Equipment. 652: 158–161. doi:10.1016/j.nima.2011.01.144

- James R. Janesick (2001). Scientific charge-coupled devices. SPIE Press. p. 4. ISBN 978-0-8194-3698-6

- Di, Dawei; Romanov, Alexander S.; Yang, Le; Richter, Johannes M.; Rivett, Jasmine P. H.; Jones, Saul; Thomas, Tudor H.; Abdi Jalebi, Mojtaba; Friend, Richard H.; Linnolahti, Mikko; Bochmann, Manfred (April 14, 2017). "High-performance light-emitting diodes based on carbene-metal-amides" (PDF). Science. 356 (6334): 159–163. doi:10.1126/science.aah4345. ISSN 0036-8075. PMID 28360136

- Pelley, J. L.; Daar, A. S.; Saner, M. A. (2009). "State of Academic Knowledge on Toxicity and Biological Fate of Quantum Dots". Toxicological Sciences. 112 (2): 276–296. doi:10.1093/toxsci/kfp188. PMC 2777075. PMID 19684286

10

Applications of Photodetectors

The varied applications of photodetectors lie in photoconductors, photomultiplier tubes, smoke detectors, compact disc players, televisions, remote controllers, microwave photonics, transmission media, etc. All these applications of photodetectors have been carefully analyzed in this chapter.

PHOTODETECTORS IN TRANSMISSION MEDIA

At the transmitting end of the optical fiber communication system, the light source is modulated with a low-frequency baseband electrical signal, and then the modulated light signal is transmitted via the optical fiber. Due to factors such as attenuation and dispersion of optical fibers, when a dimmed signal is transmitted to the receiving end, it becomes weak and has waveform distortion. The role of the optical receiver is to detect the weak light signal first, then convert it into an electrical signal, and then restore to the original baseband signal through amplification, shaping, regeneration, and decoding. Therefore, the core component of the optical receiver is a photodetector. In the entire optical fiber communication system, the photoelectric detection device mainly has two purposes: one is for terminal reception of the communication system, and the upper part of the figure is the schematic diagram of terminal receiving of the digital optical fiber communication system; the second is photoelectric conversion for the relay station. After the signal processing continues to transmit, the lower part of the figure below depicts the relay station's photoelectric conversion process.

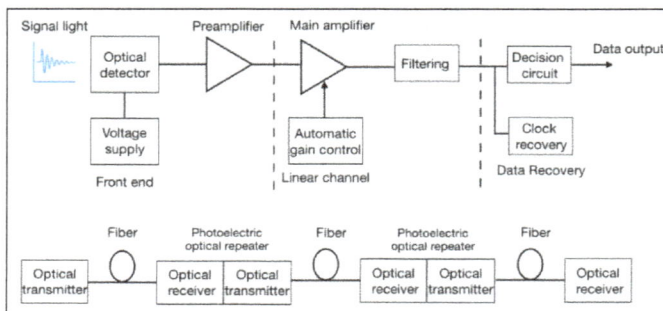

working process diagram of optical communication system.

Photoelectric Detection Principle

The photodetector is an optoelectronic device made by utilizing the photoelectric effect of a semiconductor. It converts the change of the optical signal into the change of the photocurrent, and reflects the change rule of the information. According to different conversion parameters, the semiconductor photodetector has two basic types: photoconductive type and photodiode type. The conductance of the photoconductive type detector changes with the change of luminous flux. The photodiode is always operated in the reverse bias state. It belongs to the inner photoelectric effect device, and the incident photon does not directly bombard photoelectrons, but merely raises the internal electrons from the lower energy level to the higher energy level. The differential resistance of the photodetector does not change with the luminous flux, and the generated photocurrent is proportional to the luminous flux. Both types of semiconductor photodetection devices have very fast response speeds, but have their own characteristics and different uses. In optical fiber communication systems, the most widely used photodetector is a photodiode because of the small size and long life of such detectors.

Photodetector Operating Characteristic Parameters

The main function of the photodetector is to convert the optical power signal transmitted from the optical fiber into a current signal, which carries the information of the source. Photoelectric detector basic parameters including the following 6 main features:

- Photocurrent:

  When the incident optical power of the photodetector changes, the photocurrent also changes linearly, thereby converting the optical signal into an electrical signal.

- Quantum efficiency:

  Quantum efficiency, ie, photoelectric conversion efficiency, represents the degree to which the total number of photons received by the photodetector can be converted into the total number of electrons of the photogenerated current.

- Responsiveness:

  The Responsiveness, also called photoelectric conversion sensitivity, is represented by r, which reflects how much light power is converted into photo-generated current.

- Cutoff wavelength:

  Only when the incident photon energy is greater than the bandgap of the detection device material, photogenerated carriers can be generated, forming a photocurrent. Therefore, for any photoelectric detection device made of any material, there is a minimum frequency or maximum wavelength that can be detected, i.e, the upper cutoff wavelength.

- Dark current:

The dark current represents the reverse current that occurs in the absence of light. It affects the receiver's signal-to-noise ratio and is an important quality parameter.

- Response time:

The response time (speed) indicates the ability of the photodetector to respond to the optical signal.

Fiber Optic Communication System Requirements for Photodetection Devices

The role of the photoelectric detection device is to use the photoelectric effect to convert the optical signal into an electrical signal. The main requirements for photodetection devices are: high sensitivity at the operating wavelength in order to improve the photo-electric conversion efficiency; fast response, good linearity, frequency bandwidth, and the speed of photoelectric conversion is higher than the operating speed of the system, reaching hundreds of Mbit/s. From s to thousands of Gbit/s, the bit rate of the communication system can be increased; the additional noise caused by the detection process is small, various measures are taken to reduce the internal noise of the system, and the signal to noise ratio is improved; the cost is low, the reliability is high, and the volume is small, long life, the photo-sensitive surface of the detector is matched with the core diameter of the optical fiber to improve the coupling efficiency; the operating voltage is as low as possible and easy to use. In optical fiber communication systems, the photodetector devices that meet the above requirements are the most commonly used PIN photodiodes and avalanche photodiodes (APD).

Photoelectric Detection Device

There are many materials that can be used to make photodiode detectors, such as Si, Ge, GaAs, InGaAs, GaAsP, InGaAsP, and the like. According to the photodiode detection PN junction, it can be divided into PN junction type, PIN junction type, Schottky barrier junction type, heterojunction and avalanche photodiode detectors. According to the wavelength response to light to distinguish, photodiode detector can be divided into infrared type, ultraviolet type, blue silicon type. Among them, the photodiode made of Si material, its typical peak response wavelength of 0.94 μm, its series is also more; PIN photodiode and avalanche photodiode APD's response time is short, so suitable for high-speed transmission applications; Ge The photodiode of the material is also one of the widely used optoelectronic devices. Since its band gap is smaller than

Si, it has higher sensitivity in the long wavelength band, but since the Ge material has a relatively large current, the noise is also high. InGaAs photodiodes are one to two orders of magnitude lower, so photodiodes of InGaAs compound materials are widely used. In order to meet the requirements of optoelectronic integrated circuits, integrated optical photodetectors can be fabricated using various waveguide effects.

PHOTODETECTORS FOR MICROWAVE PHOTONICS

High-speed high-power photodiodes are critical components in microwave photonics applications including radio-over-fiber, low-phase-noise microwave signal generation, antenna remoting, and arbitrary waveform generation. These applications require high power handling capability and high linearity of the photodiode (PD) to maintain high radio frequency (RF) gain and large spurious-free dynamic range (SFDR), respectively. To achieve large RF output power at microwave frequencies, several PD structures have been reported, among which the uni-traveling carrier (UTC) PD has demonstrated large bandwidth and a high saturation current. In an UTC PD, photons excite electron-hole pairs in an un-depleted absorption layer. Since only electrons are injected into the transparent drift layer, the transit time is shorter and space charge screening is reduced, compared to a conventional p-i-n PD, which has both electrons and holes in the drift region. Over the past few years, the University of Virginia (UVA) has developed InGaAs/InGaAsP/InP modified uni-traveling carrier (MUTC) PDs that achieved a high bandwidth >105 GHz, high responsivity, and low dark current. Similar to UTC photodiodes, MUTC photodiodes include an InGaAs un-depleted absorber and a depleted InP drift layer. By adding a depleted InGaAs absorber with an appropriate thickness between the un-depleted absorber and the drift layer, the responsivity can further increase. This layer also maintains a high electric field and thus facilitates electron transport at the heterojunction interface. To increase RF output power and avoid saturation, the drift layer is lightly n-type doped for charge compensation. The charge from the ionized donors pre-distorts the electric field and can counteract the space charge screening effect at a high current. A critical factor that limits the RF output power is Joule heating in the junction. Previously, it had been shown that significantly higher RF output power could be achieved by flip-chip bonding the PD onto a substrate with high-thermal-conductivity such as AlN or diamond.

High-power High-speed Photodiodes

Previously, we demonstrated back-illuminated charge-compensated MUTC PDs with 7.8 and 9.6 dBm RF output power at 110 and 100 GHz, respectively. These devices included a high-impedance 200 μm-long coplanar waveguide (CPW) between the RF pads and the PD that provided inductive peaking at 100 GHz. Recently, we flip-chip bonded similar MUTC PDs to a low-inductance transmission line (TL) on a AlN submount to increase the

RF output power beyond the 3 dB bandwidth. Using thermo-compression Au-Au bonding, the PD's bonding pads were attached to a CPW with a 130-μm signal-to-ground gap. A short tapered TL connected the stub to a 50 Ω, 50 × 154-μm2 pad with a 54-μm signal-to-ground gap and 250-μm pitch. This optimized CPW design decreased the roll-off of the frequency response beyond 100 GHz and extended the usable frequency range up to 160 GHz. Figure shows the measured RF output power and compression at 160 GHz using an optical heterodyne setup, a GGB Industries WR-6 waveguide probe with bias-T, and a VDI power meter PM5. A 9-μm diameter PD reached a maximum RF output power of −2.6 dBm at 160 GHz and −3 V, and a saturation current of 40 mA. The dark current was 0.4 nA and the responsivity was 0.2 A/W at 1550 nm. Appl. Sci. 2 μm2 pad with a 54-μm signal-to-ground gap and 250-μm pitch. This optimized CPW design decreased the roll-off of the frequency response beyond 100 GHz and extended the usable frequency range up to 160 GHz. Figure shows the measured RF output power and compression at 160 GHz using an optical heterodyne setup, a GGB Industries WR-6 waveguide probe with bias-T, and a VDI power meter PM5. A 9-μm diameter PD reached a maximum RF output power of −2.6 dBm at 160 GHz and −3 V, and a saturation current of 40 mA. The dark current was 0.4 nA and the responsivity was 0.2 A/W at 1550 nm.

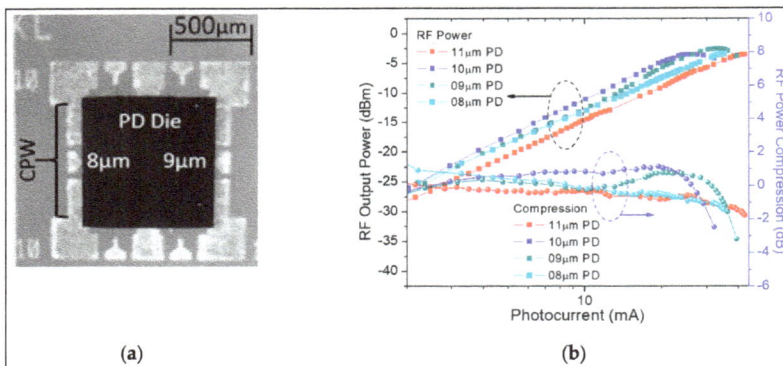

(a) Photodiode (PD) die on submount after flip-chip bonding. (b) Radio frequency (RF) output power and compression at 160 GHz of modified uni-traveling carrier photodiodes (MUTC-PDs) with different diameters.

It is well-known that a decrease in the absorption layer thickness in order to reduce the carrier transit time, results in a lower responsivity in a normal incidence photodiode. For example, in the 100-GHz PD, the absorption layer thickness was 180 nm (the drift layer thickness was 300 nm) to enable carrier transit times below 4 ps, however, the responsivity was only 0.17 A/W. To circumvent this trade-off, waveguide structures that decouple photon absorption from carrier transport can be employed. To this end, we have developed evanescently coupled wave guide integrated MUTC PDs. Figure shows the epitaxial layer structure that was grown on top of the InGaAsP input optical waveguide on the InP substrate. The InGaAs absorber includes a 100 nm thick p-type doped layer and a 100 nm-thick lightly n-type doped layer. A graded doping was designed to create a built-in electric field to support carrier transport in the un-depleted absorber. The 200 nm-thick charge-compensated InP drift layer was incorporated to reduce the junction capacitance. Intermediate-bandgap InGaAsP layers were added to prevent carrier pile-up at the band barriers. A quaternary cliff layer was designed to help increase the electric field in the depleted absorber and thus support electron transport across the hetero junction discontinuity into the drift layer.

Evanescently-coupled MUTC PDs were fabricated on a three inch InGaAsP/InP wafer that was grown by metal-organic chemical vapor deposition. The fabrication process involved a double-mesa process using dry etching and was carried out in UVA's clean room facilities.

The frequency responses of photodiodes with various areas are shown in figure. All three PDs had 3 dB bandwidths over 90 GHz and showed flat responses up to 80 GHz. The PD with an area of 24 μm^2 had a bandwidth of over 105 GHz. For applications that require efficient light detection at high modulation frequencies, the photodiode bandwidth efficiency product is an important figure of merit, which was 29 GHz, 32 GHz and 38 GHz for 24 μm^2 , 35 μm^2 , and 50 μm^2 PDs, respectively. Waveguide photodiodes fabricated from the same wafer lot achieved an RF output power of 5 dBm at 120 GHz.

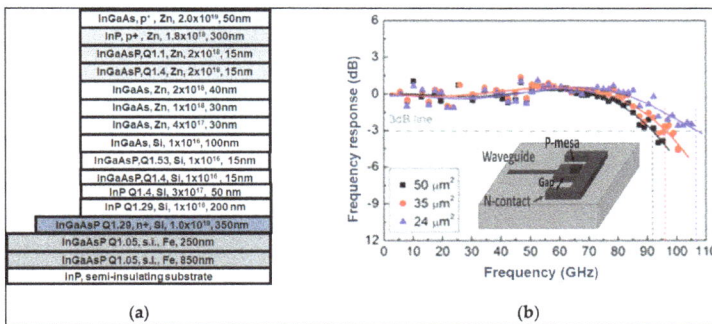

(a) Layer stack of waveguide MUTC PD; doping concentrations are given in cm^{-3}
(b) Frequency responses of PDs with different active areas; inset: schematic layout.

Integrated Photodiode-antenna Emitters

Recently, significant research has been devoted to the integration of photodiodes with antennas, or microwave photonic wireless transmitters. Due to the everlasting demand for high data rate wireless transmission, the low frequency bands are already too crowded to cope with the requirements. Higher frequency bands well into the range of millimeter-waves are needed to be deployed for next generation high data rate wireless networks. However, the hardware for the wireless data transmission at millimeter-wave frequency is far from being mature. The traditional way of implementing the hardware using electronic systems has its drawbacks, such as the high signal propagation loss inside coaxial cables, the limited bandwidth of electronics and the susceptibility to electromagnetic interference, which are incompatible with system operation at millimeter-wave frequencies. The realization of these systems using photonic techniques is advantageous, mainly due to the low-loss signal propagation in optical fibers, broad bandwidth of photonic components, and immunity to electromagnetic interference.

For a photonic wireless transmitter, the data is imposed on an optical carrier and transmitted to the antenna site through optical fiber. Then, the optical signals are converted to the electrical domain by a PD. Finally, the signals are broadcast to the users wirelessly by the antenna. The PD should maintain high speed at high optical input power in order to achieve sufficient radiated power, which is why the MUTC PD is well suited for this application.

Besides the PD performance, system integration is also an important factor. For reliable operation, the system needs to have high mechanical strength, good thermal dissipation, and carefully designed RF characteristics for optimal coupling between PD and antenna. Moreover, compact planar structures are preferred owing to their small footprint and ease of large system integration such as in phased arrays.

The effective isotropic radiated power (EIRP) is among the highest in the literature due to the high power-handling capability of the MUTC PD, and its careful integration with the antenna. V-band frequencies are of particular interest since the Federal Communications Commission allocated a 7 GHz unlicensed spectrum (57–64 GHz) for 60 GHz band communication. At 100 GHz and above, a growing number of applications including high-capacity wireless communication, non-destructive sensing/imaging, and radio astronomy continue to motivate the development of photonic components.

Photonic Emitter at 60 GHz

The figure below shows the epitaxial layer structure of the MUTC PD that was designed for 60 GHz. It includes a 400-nm-thick 1×1016 cm-3 n-type doped drift layer and an InGaAs absorbing region with a total thickness of 500 nm. The transit-time limited component of the bandwidth was estimated to be 85 GHz. The measured bandwidth and saturation power of a 10-µm diameter PD (no antenna) are shown in figure. The PD had a 3-dB bandwidth of 60 GHz while the 1-dB RF saturation power reached 16.7 and 14.3 dBm at 50 and 60 GHz, respectively.

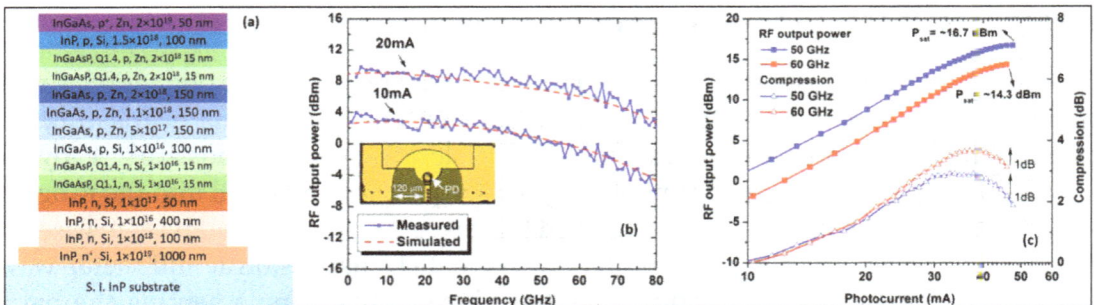

(a) Epitaxial layer design of MUTC-PD. (b) Measured bandwidth of 10-µm-diameter MUTC-PD. The nset shows a microscopic image of the fabricated device. (c) RF output power measurement of the MUTC-PD.

A coplanar patch antenna was designed to be about half a wavelength long to resonate at 60 GHz. While the CPW antenna feed was designed to be 50 Ω, a 100 µm-long high impedance (85 Ω) contact pad was integrated on the PD chip to compensate for parasitic capacitance. After the antenna was fabricated on an AlN submount, the MUTC-PD was integrated by flip-chip bonding.

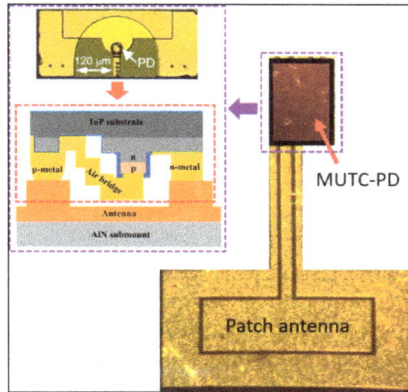

Schematic and optical image of the integrated photodiode-antenna emitter. A 50-Ω transmission line connects the patch antenna (829 μm × 2668 μm) and the flip-chip bonded MUTC PD chip.

The radiated power of the integrated photodiode-antenna emitters was measured using the setup shown in the figure. The optical RF signal was generated using an optical heterodyne setup. The radiation power was received in the far field using a commercial horn antenna and measured by a RF power meter.

The received radiation power versus frequency is shown in figure. The power at 15 mA average photocurrent amounts to −15 to −35 dBm between 50 to 75 GHz. Also shown in figure is the simulated radiation power using the circuit model shown in the inset of figure. The received RF power was calculated using Friis' Equation based on the radiated power, gain of transmitting (4.5 dBi) and receiving (15 dBi) antennas, and free-space loss (−43 dB). The simulation and experimental results agree with each other. The radiation power at 60 GHz versus the photocurrent is shown in figure. The 1-dB RF saturation power is −6.5 dBm at −5 V and 45 mA photocurrent. Using Friis' equation, the effective radiated power is 20 dBm. According to the IEEE 802.16 standard, a minimum power of −46.2 dBm is required at the receiver for a QPSK-modulated signal at 60 GHz, which should be possible once a 25-dBi receiving antenna in 15 m distance from the integrated photodiode-antenna emitter is used.

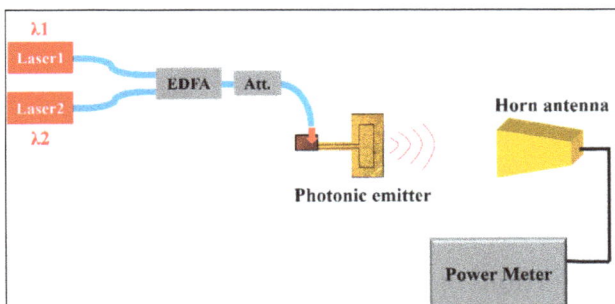

Experimental setup for measuring the radiation power at V-band. EDFA: erbium doped fiber amplifier, Att.: optical attenuator.

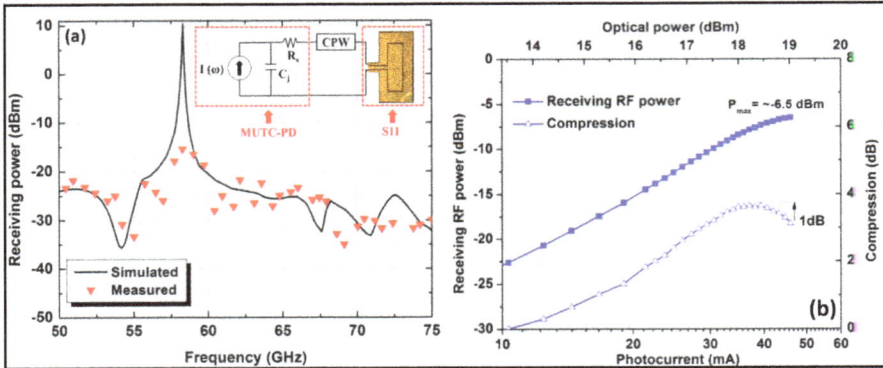

(a) Measured and simulated received RF power from the photonic emitter at 5 V bias voltage and 15 mA photocurrent. The inset shows the circuit model used to simulate the RF power. (b) RF received power versus photocurrent at 5 V bias and 60 GHz frequency.

Photonic Emitter at 100 GHz

For the W-band integrated photodiode-antenna emitter, a high-power MUTC PD with a bandwidth of 110 GHz was used. Due to the high attenuation of millimeter-wave radiation in the atmosphere, an antenna with high directional gain is desired. Commercial horn antennas and Si lenses are typically used for this purpose. However, they are bulky and incompatible with large system integration.

In order to achieve high radiation power, the transition between the PD and the antenna needs to be carefully designed. It has been shown that impedance matching has a significant effect on the radiation power of integrated photonic emitters and that maximum RF power can be extracted from the PD by conjugate impedance matching. In, a matching network was used to achieve conjugate matching between the MUTC PD and the Vivaldi antenna.

The figure below shows the integrated photonic emitter after the PD die was flip-chip bonded onto the antenna with a matching network on AlN.

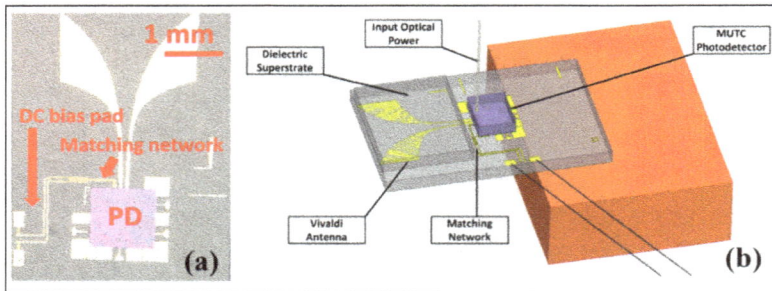

(a) Optical image and (b) 3-D schematic of the integrated photonic emitter.

In order to suppress distortions of the radiation pattern due to substrate modes, an AlN superstrate was placed on top of the antenna. The scattering parameter S11 of 5-, 6-, and 14-μm-diameter PDs were measured and are shown in the figure after de-embedding the RF pads. Also shown in figure is the S11 of the Vivaldi antenna with a matching network. Owing to the optimized matching network, the input impedance of the Vivaldi antenna is transformed close to the conjugate impedance of the 5- and 6-μm-diameter PDs, while it is far from the conjugate matching condition of the 14-μm-diameter PD.

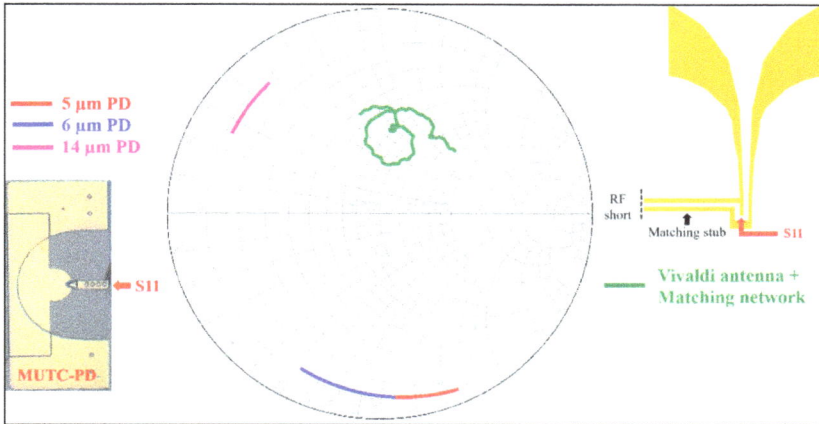

Scattering parameters of 5-, 6-, and 14-μm-diameter PDs and the Vivaldi antenna with a matching network.

The E-plane radiation power of the integrated photonic antenna was characterized in the far field using the setup shown in the figure.

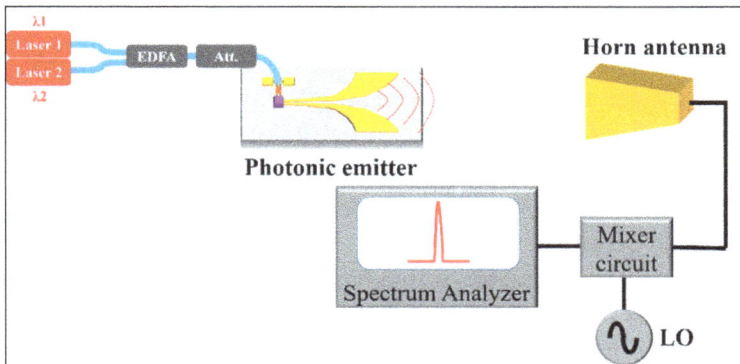

Experimental setup for measuring the radiation power of the photonic emitter at W-band. LO: local oscillator. The distance between photonic emitter and horn antenna was 60 cm.

Devices with 5-, 6-, and 14-μm-diameter PDs were measured and their EIRP from 95 to 110 GHz are shown in the figure. Photonic emitters with 5- and 6- μm-diameter PDs reached higher EIRP than the photonic emitter with the 14-μm-diameter PD, which can be explained by the larger PD junction capacitance and poor impedance matching. The photonic emitter with 5-μm-diameter PD reached 5 dBm at 110 GHz, and the -6-dB bandwidth for all three photonic emitters was 10 GHz. The fact that the EIRP varies across the measured frequency range might result from the varying impedance matching condition. The EIRP at 100 GHz as a function of photocurrent is shown in the figure.

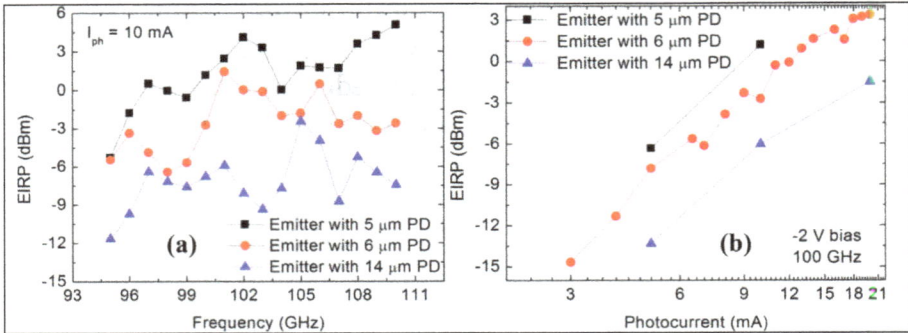

(a) Effective isotropic radiated power (EIRP) of the integrated photonic emitters with 5-, 6-, and 14-µm-diameter PDs at 10-mA photocurrent from 95 to 110 GHz. (b) EIRP at 100 GHz as a function of average photocurrent.

High-power Photodiodes on Si

The development of photonic integrated circuits on silicon has large technological and commercial significance, since it can leverage the mature Si CMOS technology to reduce the manufacturing costs. However, in order to achieve optoelectronic functionality at 1.55 µm wavelength, typically other materials have to be integrated on silicon. For PDs, the candidates include group III-V semiconductors and Germanium.

Heterogenous Photodiodes

Heterogeneous silicon photonics, i.e the integration of group III-V materials onto silicon, benefits from the mature Si processing technology, while fully exploiting the high-performance of III-V materials. Since bandgap-engineering is available in III-V semiconductors, complex photodiode heterostructures can be designed, which have been shown to enable high-power high-linearity analog applications. To date, three approaches have been reported for the heterogeneous integration of III-V photodiodes on silicon: (i) molecular bonding, (ii) adhesive bonding, and (iii) III-V material growth on Si. Previously, it has been demonstrated that molecular die and wafer bonding are technologies that combine different materials without compromising their properties. These approaches have produced high-performance photodiodes with a low dark current, high responsivity at 1.55 µm wavelength, high speed, and high power. In waveguide MUTC PDs were fabricated from an InGaAs/InP die that was wafer-bonded onto silicon-on-insulator (SOI). The PDs had an internal responsivity of 0.95 A/W, with a very low dark current of 10 nA (2.9 mA/cm2), a bandwidth of 48 GHz, and high RF output power of 12 dBm at 40 GHz. this work was extended to waveguide photodiodes with a bandwidth of 65 GHz. Figure shows the device structure and the measured frequency responses, respectively. Unlike some earlier work, an inverted photodiode layer stack was used, which resulted in the p-contact being on top after dye bonding. To enable efficient optical coupling from the silicon waveguide through the low-index InP drift layer into the absorber, the Si waveguide was designed to be 300 nm wide (inset of figure). It follows that the optical mode is no longer confined. Instead, it is pushed upwards into the active photodiode region. It should be noted, that the heterogeneous integration process, which allows changing the widths of the PD mesa and the Si waveguide independently, enabled this design. Responsivity and dark current were 0.84 A/W at 1.55 μm and 1 nA at −3 V, respectively. The output power was −2 dBm at 70 GHz for a PD with an area of 75 μm.

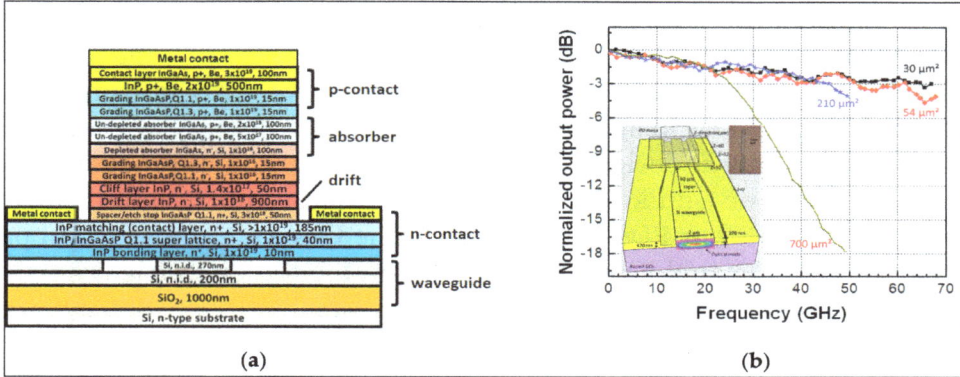

(a) Device structure of heterogeneous MUTC PD on silicon-on-insulator (SOI); (b) Measured frequency responses for PDs with different area. The inset shows a schematic view of the PD.

Ge-on-Si Photodiode Arrays

For monolithic integration, germanium-on-silicon has been widely investigated for near-infrared optoelectronics, and despite a lattice mismatch of 4%, Ge heterogeneous epitaxy on Si substrates has been successfully demonstrated with sufficiently low defect densities. Compared to III-V growth on Si, Ge-on-Si is more compatible with the Si CMOS process. In addition, Ge can be directly grown on Si without buffer layers, which makes optical coupling between passive structures and Ge photodetectors easier. As a result, Ge-on-Si has become a well-developed platform for photonic integrated circuits. Today, many foundries offer complete suites of technologies based on this platform, including the American Institute for manufacturing Integrated Photonics (AIM Photonics), IMEC, GlobalFoundries, Institute of Microelectronics (IME), STMicroelectronics, Taiwan Semiconductor Manufacturing Company (TSMC), and TowerJazz.

Ge-on-Si PDs with a low dark current, large bandwidth and high responsivity are well-developed in monolithic Si photonic platforms for digital applications, and large scale digital systems have been realized. However, only a few Ge-on-Si PDs for high-power analog applications have been demonstrated. In addition to high responsivity and large bandwidth, PDs for microwave photonics applications should possess high-power handling capability and high linearity in order to achieve large RF gain and high spur-free dynamic range. For example, in order to achieve an RF link gain of 0 dB with a modulator with V_π of 3 V, the photocurrent should be as high as 27 mA for an intensity-modulated direct detection link and 14 mA for a phase-modulated link. These photocurrents are much larger than the ones that typically occur in digital applications. The fact that the PDs need to maintain a large bandwidth and high linearity at these levels of photocurrents, puts stringent requirements on PDs in Si photonics.

Lack of material versatility limits the design freedom of epi-layers in monolithic Ge-on-Si PDs. To mitigate this drawback, PD arrays have been proposed. In 8 Ge PDs were connected in a traveling-wave fashion and the optical signal was fed from both sides of the active region in order to enhance the power handling capability. Four Ge PDs were connected in parallel in a compact array to enhance the RF output power. The optical image and schematic of the PD array are shown in figure. The device layout was designed using the AIM Si photonics Process Design Kit (PDK. Three optical waveguide Y-junctions were used to equally split the input optical power to feed the four PDs. The outputs of the PDs were connected in parallel to the shared ground-signal-ground (GSG) pad on the top dielectric layer through vias. The device was fabricated on a multi project wafer run in AIM's Si photonics foundry.

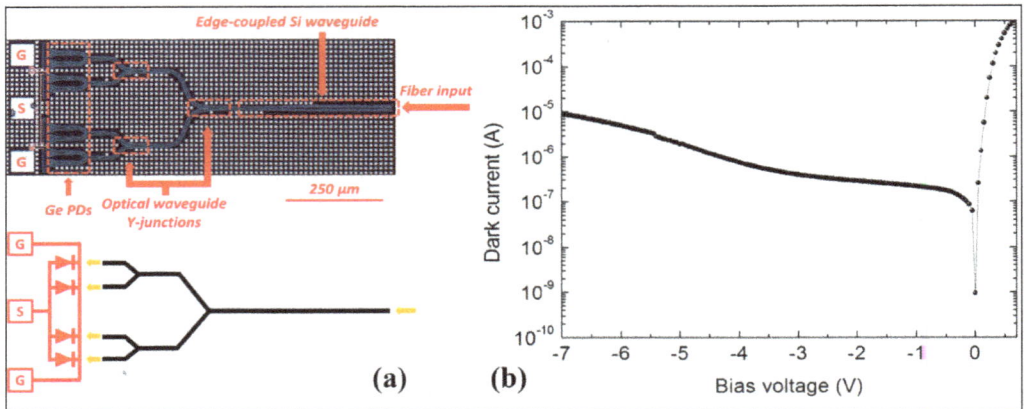

(a) Optical image and schematic of the PD array, and (b) I-V characteristics.

The I-V characteristics of the PD array are shown in the figure. The dark current is as low as 0.3 µA at −2 V and 1 µA at −5 V bias. The PD array has a high external responsivity of 0.58 A/W at 1550 nm wavelength (no anti-reflection coating at the waveguide edge coupler) when measured with a lensed fiber with 3 µm spot size. The fiber-chip coupling loss was estimated to be 1.7 dB.

The frequency response of the PD array was measured at −5 V bias at various photocurrents from 1 mA to 20 mA. The results are shown in figure. The 3-dB bandwidth is 15 GHz up to a photocurrent of 15 mA and remains as high as 12 GHz at 20 mA photocurrent. The RF output power of the PD array was measured at the 3-dB bandwidth frequency under different reverse voltages. As shown in figure, the RF power increases with reverse voltage and begins to saturate above −5 V bias. The RF power and the saturation current at 1-dB compression are 7 dBm and 19 mA, respectively, under −7 V at 15 GHz.

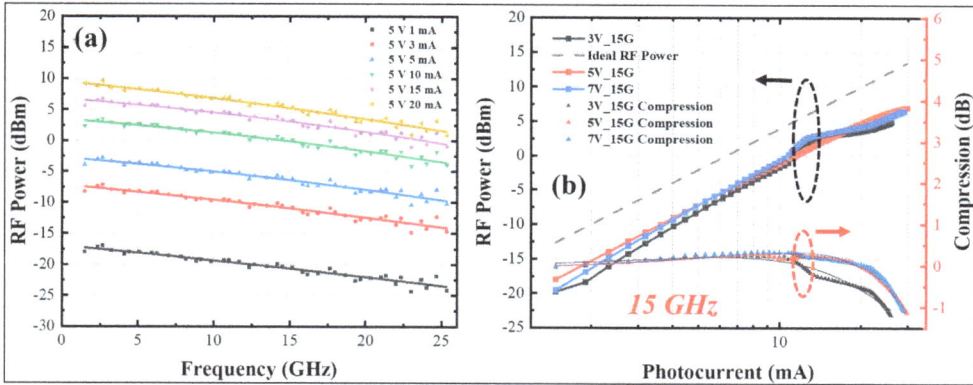

(a) Frequency responses of PD array at different photocurrents.
(b) RF output power and compression of the PD array at 15 GHz.

In order to demonstrate the applicability of a Ge-on-Si PD array in a microwave photonics application, a balanced PD pair integrated with a Mach-Zehnder delay line interferometer (MZ DLI) was used to demodulate and detect a phase-modulated signal in an analog photonic link. The optical image and the schematic of the integrated receiver are shown in the figure. The optical input signal is fed through one of the vertical grating couplers and then split by a Y-junction. One arm includes a 1.4 mm-long silicon waveguide to delay the optical signal by 25 ps to allow for interferometric demodulation of the phase modulated signal. The demodulator worked at the quadrature point by adjusting the laser wavelength.

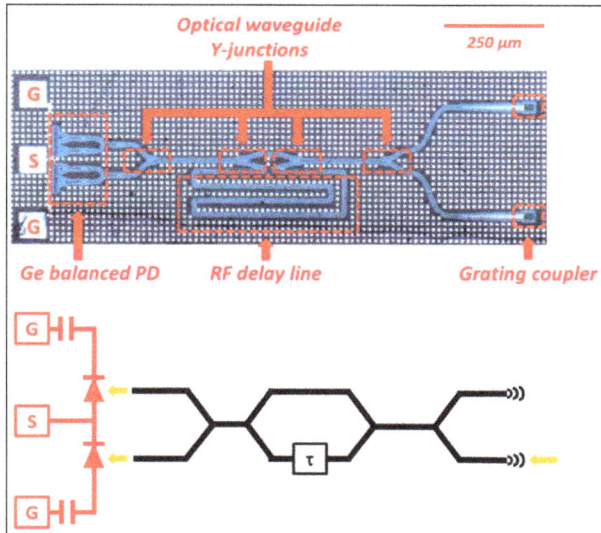

Optical image and schematic of the interferometric demodulator.

The RF gain of the phase-modulated link was measured using the setup shown in figure. A commercial optical phase modulator with a V_π of 7 V was driven by a signal generator to modulate the phase of the optical carrier from the 1550 nm laser. The demodulated signal was recorded from a spectrum analyzer. The RF link gain versus frequency at different photocurrents is shown in figure. Also shown in the figure is the calculated link gain versus frequency according to. The measurement agrees well with the theory once the 3-dB bandwidth of the balanced PD pair of 20 GHz was taken into account (blue curve). At 7 mA photocurrent per PD, the difference between the measurement (black curve) and the ideal model (red curve) increases, which indicates a reduction of the bandwidth of the balanced PDs due to saturation. A higher RF gain can be expected from traveling wave balanced PD arrays.

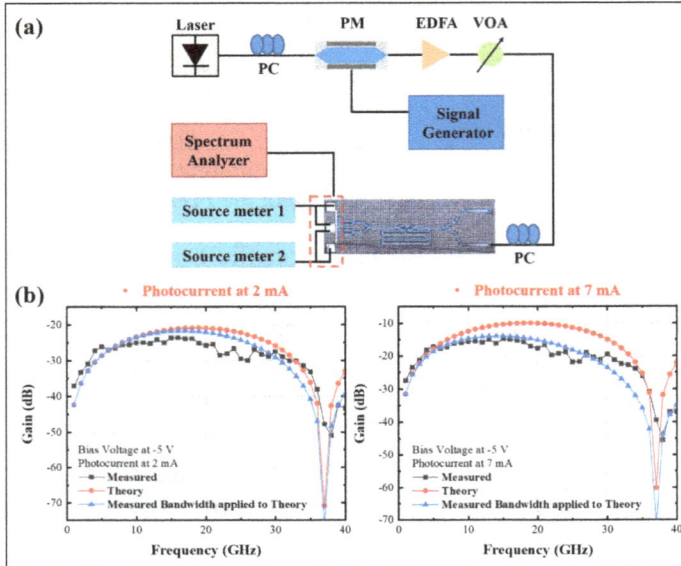

(a) Phase-modulated link gain measurement setup. (b) Measured and calculated RF gain spectra of the phase-modulated link at 2 and 7 mA photocurrent for each PD.

RESONANT-CAVITY-ENHANCED PHOTODETECTORS

Resonant-cavity-enhanced photo detectors (or, RCE photodetectors) enable improved performance over their predecessors by placing the active device structure inside a Fabry–Pérot resonant cavity. Though the active device structure of the RCE detectors remains close to other conventional photodetectors, the effect of the optical cavity, which allows wavelength selectivity and an enhancement of the optical field due to resonance, allows the photo detectors to be made thinner and therefore faster, while simultaneously increasing the quantum efficiency at the resonant wavelengths.

Advantages

The quantum efficiency of conventional detectors is dominated by the optical absorption (electromagnetic radiation) of the semiconductor material. For semiconductors with low absorption coefficients, a thicker absorption regions is required to achieve higher quantum efficiency, but at the cost of the signal-processing bandwidth of photodetectors.

A RCE detector improves the bandwidth significantly. The constructive interference of a Fabry–Perot cavity enhances the optical field inside the photodetector at the resonance wavelengths to achieve a quantum efficiency of close to unity. Moreover, the optical cavity makes the RCE detectors wavelength selective. This makes RCE photodetectors attractive for low crosstalk wavelength demultiplexing. Improved quantum efficiency gives less power consumption. Higher bandwidth gives faster operation.

The RCE photodetectors have both wavelength selectivity and high speed response making them ideal for wavelength division multiplexing applications. Optical modulators situated in an optical cavity require fewer quantum wells to absorb the same fraction of the incident light, and can therefore operate at lower voltages. In the case of emitters, the cavity modifies the spontaneous emission of light-emitting diodes (LED) improving their spectral purity and directivity.

Thus optical communication systems can perform much faster, with more bandwidth and can become more reliable. Camera sensors could give more resolutions, better contrast ratios and less distortion. For these reasons, RCE devices can be expected to play a growing role in optoelectronics over the coming years.

Theory of RCE Photo Detectors

The RCE photo detectors can provide:

- Higher quantum efficiency,

- Higher detection speed,

- Wavelength selective detection, than compare to a conventional photodiode.

Quantum Efficiency of RCE Photodetectors

The RCE photodetectors are expected to have higher quantum efficiency η than compare to conventional photodiodes. The formulation of η for RCE devices gives insight to the design criteria.

A generalized RCE photodetector schematic as given in figure can give the required theoretical model of photodetection. A thin absorption region of thickness d is sandwiched between two relatively less absorbing region, substrate, of thickness L_1 and L_2. The optical cavity is formed by a period of $\lambda/4$ distributed Bragg reflector (DBR), made of non-absorbing larger bandgap materials, at the both end of the substrate. The front mirror has a transmittance of t_1 and generally has lower reflectivity than compare to the mirror at back ($R_1 < R_2$). Transmittance t_1 allows light to enter into the cavity, and reflectivity R_1 ($=r_1^2$) and R_2 ($=r_2^2$) provides the optical confinement in the cavity.

The active region and the substrate region have absorption coefficient α and α_{ex} respectively. The field reflection coefficients of the front and the back mirrors are $r_1 e^{-j\phi_1}$ and $r_2 e^{-j\phi_2} r$ respectively, where ϕ_1 and ϕ_2 are the phase shifts due to the light penetration into the mirrors.

The optical microcavity allows building up an optical field inside the optical cavity. In compare to conventional detector, where light is absorbed in a single pass through the absorption region, for RCE detectors trapped light is absorbed each time it traverses through the absorption region.

The Quantum efficiency η for a RCE detector is given by:

$$\eta = (1-R_1)(1-e^{-\alpha d})[\frac{(e^{-\alpha_{ex}L_1} + r_2^2 e^{-\alpha_{ex}L_2 - \alpha_c L})}{1 - 2r_1 r_2 e^{-\alpha_c L}\cos(2\beta L + \phi_1 + \phi_2) + (r_1 r_2)^2 e^{-\alpha_c L}}]$$

Here $\alpha_c = (\alpha_{ex}L_1 + \alpha_{ex}L_2 + \alpha d)/L$. In practical detector design $\alpha_{ex} << \alpha$, so α_{ex} can be neglected and η can be given as:

$$\eta = (1-R_1)(1-e^{-\alpha d})[\frac{(1+R_2 e^{-\alpha d})}{1 - 2\sqrt{R_1 R_2}e^{-\alpha_c d}\cos(2\beta L + \phi_1 + \phi_2) + (R_1 R_2)e^{-\alpha_c d}}]$$

The term inside the [] represents the cavity enhancement effect. This is a periodic function of $2\beta L + \phi_1 + \phi_2$, which has minima at $2\beta L + \phi_1 + \phi_2 = 2m\pi$. And η enhanced periodically at resonance wavelength that meets this condition. The spacing of the resonant wavelength is given by the Free Spectral Range of the cavity.

The peak value of η at resonant wavelength is given as:

$$\eta = (1-R_1)(1-e^{-\alpha d})[\frac{(1+R_2 e^{-\alpha d})}{(1-\sqrt{R_1 R_2}e^{-\alpha_c d})^2}]$$

for a thin active layer as $\alpha d << 1$, η becomes:

$$\eta = (1-R_1)\alpha d[\frac{(1+R_2 e^{-\alpha d})}{(1-\sqrt{R_1 R_2}e^{-\alpha_c d})^2}]$$

This is a significant improve from the quantum efficiency of a conventional photodetector which is given by:

$$\eta = (1-R)\alpha L.$$

This shows that higher quantum efficiency can be achieved for smaller absorption region.

The critical design requirements are : a very high back mirror reflectivity and a moderate absorption layer thickness. At optical frequencies metal mirrors have low reflectivity (94%) when used on materials like GaAs. This makes metal mirrors inefficient for RCE detection. Whereas distributed Bragg reflector (DBR) can provide reflectivity near unity, are ideal choices for RCE structures.

For a R2=0.99 and α=10⁴ cm-1 with a R1=0.2 a η of 0.99 or more can be achievable for d=0.7–0.95 μm. Similarly for different values of R1 very high η is possible to achieve. However, R1 =0 limits the length of thickness region, d>5 μm can achieve 0.99 η but at the cost of bandwidth.

Detection Speed of RCE Photodiodes

The detection speed depends upon the drift velocities of the electrons and holes. And between these two holes have slower drift velocity than the electrons. The transit time limited bandwidth of conventional p-i-n photodiode is given by:

$$f_{transit} = 0.45 \frac{vh}{L}$$

However the quantum efficiency is a function of L as:

$$\eta = (1 - R)\alpha L.$$

For a high speed detector for a small value of L, as α is very small, η becomes very small (η<<1). This shows for an optimum value of quantum efficiency the bandwidth has to sacrifice.

A p-i-n RCE photodetector can reduce the absorption region to a much smaller scale. In this case the carriers need to traverse a smaller distance as well, L_1 (< L) and L_2 (< L) for electrons and holes respectively.

The length of L1 and L2 can also be optimized to match the delay between the hole and electron drift. And the transition bandwidth becomes:

$$f_{transit} = 0.45 \frac{vh + v_e}{L + d}$$

As in most of semiconductors v_e is more than v_h the bandwidth increases drastically.

It's been reported that for a large device of L=0.5μm 64 GHz of bandwidth can be achieved and a small device of L=0.25μm can give 120 GHz bandwidth, where conventional photodetectors have bandwidth of 10–30 GHz.

Wavelength Selectivity of RCE Photo Detectors

A RCE structure can make the detector wavelength selective to an extent due to the resonance properties of the cavity. The resonance condition of the cavity is given as $2\beta L + \phi_1 + \phi_2 = 2m\pi$. For any other value the efficiency η reduces from its maximum value, and vanishes when $2\beta L + \phi_1 + \phi_2 = (2m+1)\pi$. The wavelength spacing of the maxima of η are separated by the Free Spectral Range of the cavity, given as:

$$FSR \frac{\lambda^2}{2n_{eff}(L + L_{eff,1} + L_{eff,2})}$$

Where neff is the effective refractive index and Leff, iare the effective optical path lengths of the mirrors.

Finesse, the ratio of the FSR to the FWHM at the resonant wavelength, gives the wavelength selectivity of the cavity.

$$finesse = \frac{\pi(R_1 R_2)^{1/4} e^{-\frac{\alpha d}{2}}}{(1 - \sqrt{R_1 R_2} e^{-\alpha_c d})^2}$$

This shows that the wavelength selectivity increases with higher reflectivity and smaller values of L.

Material Requirements for RCE Devices

The estimated superior performance of the RCE devices critically depends on the realization of very low loss active region. This enforces the conditions that: the mirror and the cavity materials must be non-absorbing at the detection wavelength; and the mirror should have very high reflectivity so that it gives highest optical confinement inside the cavity.

The absorption in the cavity can be limited by making the bandgap of the active region smaller than the cavity and the mirror. But a large difference in the bandgap would be a blockage in extraction of photo generated carriers from a heterojunction. Usually a moderate offset is kept within the absorption spectrum.

Different material combinations satisfy all of the above criteria and are therefore usable to the RCE scheme. Some material combinations used for RCE detection are:

- GaAs(M,C)/AlGaAs(M)/InGaAs(A) near 830-920nm.

- InP(C)/In$_{0.53}$Ga$_{0.47}$As(M)/In$_{0.52}$Al$_{0.48}$As(M)/In$_{0.53-0.7}$GaAs(A) near 1550nm.

- GaAs(M,C)/AlAs(M)/Ge(A) near 830-920nm.

- Si(M,C)/SiGe(M)/Ge(A) near 1550nm.

- GaP(M)/AlP(M)/Si(A,S) near visible region.

Future of RCE Photodiodes

There are many examples of RCE devices, like p-i-n photodiode, avalanche photodiode, schottky diode are made that verifies the theory successfully. Some of them are in use in practical purposes as well as there is a future prospect in use as modulators, optical logics in wavelength division multiplexing (WDM) systems which could enhance the quantum efficiency, operating bandwidth, wavelength selectivity.

RCE detectors are preferable in potential price and performance in commercial WDM systems. RCE detectors have very good potential for implementations in WDM systems and improve the performance significantly. There are various implementations of RCE modulators are made and there is a huge scope for further improvement in performance of those. Other than the photodetectors the RCE structures have lots of other implementations and a very high potential for improved performance. A Light Emitting Diode (LED) can be made to have narrower spectrum and higher directivity to allow more coupling to optical fibre and better utilization of the fibre bandwidth. Optical amplifiers can be made to have more compact, thus lower power required to pump and also at lower cost. Photonic logics will work more efficiently than they do. There will be much less crosstalk, more speed, more gain with simple design.

OTHER APPLICATIONS OF PHOTODETECTORS

- Photo detectors are used in various different applications such as radiation detection, smoke detection, flame detection and to switch on relays for street lighting.

- The circuits that use photodiodes use either normally closed or normally open contacts depending on the desired operation.

- In a smoke detector circuit the photo diode is attached to a relay switch, this switch is normally closed and attached to the fire alarm. When the photo diode conducts it picks up the relay switch, this causes the normally closed switch to open preventing the alarm from activating. When the photo diode fails to conduct, the normally closed contact activates the alarm.

- Photo diodes are also used in modern oil burning furnaces as a safety feature. The photo diode is comprised of lead sulphide and is used to detect the flame from the boiler, in the event that the flame goes out or fails to occur the photo diode opens the circuit, cutting power to the motor and step up transformer.

- Another commonly used application is street lights. The photo diode in the circuit uses switch-on relays to turn on the street lights when the diode fails to conduct and turns the lights off with when sufficient light is present.

- Another application is the AFM (Atomic Force Microscope), a laser beam is projected from a laser diode onto the back of the cantilever, and the beam is then reflected to a photodiode. The position of the beam of light on the diode gives the (x,y,z) position of the material as the probes of the cantilever scraps across the surface of the material. This gives a three dimensional representation of the surface being scanned.

- Photodiodes are also used with lasers to form security system. When the light projected by a laser to the photodiode is broken a security alarm is tripped.

Permissions

Index

www.ingramcontent.com/pod-product-compliance
Lightning Source LLC
Chambersburg PA
CBHW080240230326
41458CB00096B/2739